THE
FARMER'S
DOG

JOHN HOLMES

POPULAR DOGS
London

Popular Dogs Publishing Co. Ltd
Random House, 20 Vauxhall Bridge Road, London SW1V 2SA

Random House Australia (Pty) Limited
20 Alfred Street, Milsons Point, Sydney, New South Wales 2061, Australia

Random House New Zealand Limited
18 Poland Road, Glenfield, Auckland 10, New Zealand

Random House South Africa (Pty) Limited
Isle of Houghton, Corner of Boundary Road & Carse O'Gowrie
Houghton 2198, South Africa

Distributed in North America by
Diamond Farm Book Publishers, Division of
Yesteryear Toys and Books Inc.
RR3, Brighton, Ontario, Canada, KOK 1HO,
Box 537, Alexandria Bay, NY 13607, USA

First published 1960
Revised edition 1963
Reprinted 1966
Revised edition 1970
Reprinted 1973
Revised edition 1975
Reprinted 1976, 1978, 1982
Revised 1976, 1978, 1982
Revised edition 1984
Reprinted 1986, 1988, 1989, 1991, 1992, 1994, 1995, 1997, 2000, 2001, 2004, 2009

© John Holmes 1960, 1963, 1970, 1973, 1975, 1984

The Random House Group Limited Reg. No. 954009
www.randomhouse.co.uk

Set in Baskerville by BookEns, Saffron Walden, Essex

ISBN 978 0 09 156121 5

Penguin Random House is committed to a sustainable future for
our business, our readers and our planet. This book is made from
Forest Stewardship Council® certified paper.

MIX
Paper | Supporting
responsible forestry
FSC® C018179

Printed and bound in Great Britain by Clays Ltd, Elcograf S.p.A.

CONTENTS

ILLUSTRATIONS

All the photographs are of the author's own dogs

AUTHOR'S INTRODUCTION

My original intention was to scrap the old introductions and write an entirely new one for this edition. When I read them over, however, I felt that the introductions to the earlier editions were worth reprinting as they stand.

When I was seventeen I tied for first place in a Young Farmers' Club essay competition. The subject: 'A profitable sideline to agriculture'; and my sideline: dog breeding. But if anyone had told me then that, one day, I should be earning my living from dogs I should have refused to believe them. Why I should have been so keen on dogs I do not know, as I had no encouragement from any direction. Whereas handling horses came almost as second nature to me, I had to learn all I wanted to know about dogs.

Advice was usually hard to find and often I wished I could buy a good book on working dogs. But there were no books – good or bad. That is one of the reasons why I have written this. In doing so I have tried to take my mind back to my early days in an effort to explain the things that then troubled me. If, at times, I appear to talk down to my readers I hope that they will forgive me. This is because I am writing for the person who has had no practical experience of working dogs. In teaching dogs – or people – I firmly believe that a sound knowledge of the fundamentals is far more useful than a smattering of the whole subject.

Of course, I did get much valuable advice from men who not only understood dogs but also realized their value. Men like old Jock, whose weakness for the bottle had changed him from a very successful sheep farmer to a somewhat nomadic shepherd taking seasonal jobs in various places. He frequently worked for my father and, as Jock had a great understanding of all animals, we were great friends.

One day, when the farm road was covered with packed frozen snow, I found him more or less 'swimming' on top of it. He could not get on his feet, and his face and hands were bleeding. Fussing round him, licking his wounds, was a little black collie bitch. I helped him to his feet on the grass verge where he clung desperately to a wire fence. Having apologized, as he always did, for being 'foo', he soon forgot his own sorry state – or was unable to appreciate it! 'Wha'd' ye think o' ma wee bitch, Jack?' he said, clinging to the top strand with one hand and with the other pointing vaguely in the bitch's direction.

'She looks all right,' I replied. 'What's she like as a worker?'

'She–e's a topper,' he replied emphatically. 'The best worker in Pairthshire.'

'Ah, but you haven't seen my young bitch Jean working yet,' I replied.

This really excited Jock who replied with enthusiasm, 'C'mon an' we'll ha'e a trial.'

'A trial!' I said. 'You're not fit to have a trial.'

Swaying precariously with one hand still on the fence and indicating the bitch with the other, he replied, amid snorts of indignation: 'Huh! There's naethin' wrang wi' the *dug*.'

Having lived and worked with men of that sort who realized the value of a good dog – under any circumstances – I then went to that 'foreign' country, England, having taken a foreman's job in Kent. My first letter home contained the extraordinary information that I had seen, with my own eyes, men running after sheep and barking like dogs! Strangely enough there was a dog on this farm exempted from licence duty as a worker. But if he were let out it meant all the more running for the men.

Since then I have seen a lot of strange and amusing sights, the most recent being one man in a jeep, assisted by five men on foot, driving six heifers along a country lane. In these days of high wages and work study on the farm such sights strike me as more than slightly ludicrous.

Not that I can tell you how many man hours a dog can save. Nor do I believe that anyone can tell you that. Variations in men, dogs and work to be done make such calculations impossible. Neither do I expect to persuade anyone to buy a dog, as those who don't want one are unlikely to buy a book entitled *The Farmer's Dog*. Only a minority of dog owners, however, use their dogs to anything like full advantage. Few have any idea of their own dog's capabilities – or incapabilities!

My first object is to help readers to find a dog that can be trained. My second is to help them to train it and to realize what it can be taught. I make no attempt to cater for those who have ambitions to work dogs in trials. It is the general farmer, and especially the young farmer, with limited time and facilities for training that I have in mind. And I have written quite a lot in this book about simply looking after the farm dog. This I do as much for the sake of the dogs as for the benefit of my readers. I was brought up in the old school where sentimentality was regarded as a sign of weakness and I do not think anyone can call me a sentimentalist. Sentimental cranks are one of the bugbears of my life. But, as one born and brought up to farming, I am thoroughly ashamed of the conditions under which many farm dogs are kept. If this book does nothing more than help to improve some of those conditions it will have served a good purpose.

To me, training dogs is easy compared to writing a book on the subject. Things which all happen at the same time have to be given separate chapters. Sorting them out into their right order gives me as much thought as writing the book – and then I am never sure what *is* the right sequence! I therefore strongly advise reading a book of this sort as a whole, not in bits and pieces.

I have been told that articles which I write are typified by the frequence of 'in my opinion' and 'I believe'. But editors of books dislike repetition of such phrases. I should, therefore, like to point out that the whole of this book *is my opinion*, and that I do not think it is worth any more than many other opinions. There are probably others better qualified than I am to write a book on this subject. I only wish they would!

If a way of doing anything were definitely the best then

progress would have stopped. The methods I describe are the best I happen to know at the time of writing. The only claim I can make which perhaps some writers cannot, is that I have tried out the methods or seen them tried personally. If I were to revise this book in a few years' time it is quite possible that I should be able to improve on the methods described here. Indeed, I should hope to be able to make some improvements. In the same way it is quite possible that a reader describing my methods to an experienced trainer may be told of better ones.

If I appear to be labouring this point it is because I meet so many people who accept as infallible all that they read in books or newspapers, hear on the radio or see on TV. The most one can hope from any instructive book is that it contains the honest opinions of the author and that the author knows what he is writing about.

1960   J.H.

Revising a book for a new edition is, naturally, a satisfying task but, for me, it always has its sad moments. Many of the dogs to which I refer in the present tense and whose pictures illustrate the book are, inevitably, no longer with me. In this edition I have also had to change from present to past tense my reference to a very good friend and adviser. That grand old man of sheepdogs David Dickson passed away – as he must surely have wished, while actually working his dog at a trial.

Since *The Farmer's Dog* first appeared I have had letters from readers all over the world. Naturally I am very pleased about this – and must admit to being a little surprised at how well the book has sold abroad. What has surprised me even more, and pleased me just as much, is the number of readers who have no intention of ever training a sheepdog but who simply want to know how one is trained.

None of my correspondents suggests any omissions or shortcomings so I have not found it necessary to make many alterations. The only real changes concern castration and spaying; these are merely due to more evidence in support of the views I already held and not to any change of opinion.

1963   J.H.

One of the prophecies I made in the first introduction has been proved wrong. After twenty-three years I do not have any really new ideas on training to offer. This may, partly, be due to the fact that I had trained many more dogs before I wrote the book than I have done since. But I think it is mainly because the dog itself has changed very little. My chief alteration has been in writing a completely new chapter on feeding. Not because the dog's digestive system has changed, but because the availability of various foods has changed and knowledge of the dog's food requirements has improved.

Several books have now been written about sheepdogs and their training. I think it is a very good idea for the beginner to read as many as possible, think carefully about what he has read and form his own opinion. We live in an age of teach-ins, seminars, clinics and what have you, which makes it much easier to obtain first hand, practical advice, and this applies to sheepdog training as much as to anything else. Thanks, largely, to the highly successful TV series 'One Man and His Dog', sheepdogs and their training have suddenly become big business. Regular auction sales are held in Scotland, England and Wales when good young dogs make prices quite unheard of when this book was first written. It is interesting to note that many of the dogs go to farmers who have no intention of running them in trials.

On the other hand many people, having seen what a good dog can be taught to do, are inspired to have a go themselves. These people are catered for by the ever increasing number of novice trials held during the winter months. Apart from giving beginners the opportunity of comparing their dogs with those belonging to other people, these trials are nearly always run by experienced trainers who are usually only too pleased to give helpful advice. All of which has helped to improve the general standard of training.

In the introduction to the first edition I emphasized that I

have no intention of teaching people how to win trials and so I
have been delighted (and not a little surprised) at the number
of people who have told me that they have trained their dog to
novice trial standard entirely from this book.

It is a pity so many general farmers still do not seem to
appreciate the value of a good dog – or buy a good dog and ruin
it! They go to endless trouble to learn to operate a new
machine but they make no effort to learn how to work a dog.
But the dog may well save far more man hours than the
machine and, given the opportunity, will be a faithful companion at the same time.

1983   J.H.

Much has changed since 1960 when I wrote the first introduction to this book.

Who would have believed that over 240 Border Collies
would be entered for Crufts dog show in 1994? And that is not
counting all those in the Obedience and Agility competitions.
I have no doubt that the vast majority of these dogs have in-
herited a strong herding instinct and therefore would work –
how they would work is another matter. While many thousands more Border Collies are registered today than thirty
years ago it seems to be more difficult than ever to find sensible
dogs that will get on and do the job for which they are needed.
It is for this reason that some people are trying New Zealand
Huntaways, Australian Kelpies and Australian Cattle Dogs.

Perhaps the most significant change to the dog scene in recent years has been the quest for knowledge with ever increasing opportunities to obtain it. Apart from numerous
books on the subject it is now possible for the novice to attend
stock dog clinics and teach-ins of one sort of another, where
expert first-hand advice can be obtained. Several Border
Collie Clubs, the Agricultural Training Board and a number of
individuals run training classes for herding breeds.

In Canada and America stock dog clinics for all herding
breeds are extremely popular. They are also to be found in
Australia and no doubt many other countries. I strongly advise
novices to take advantage of these opportunities.

1994   J.H.

Part I

THEORY

I

Instincts

Hunting and Herding – The Pack Instinct
Other Instincts – Development and Balance of Instincts

HUNTING AND HERDING

WHETHER it be a dog, a cow or a combine harvester, the first essential to maximum efficiency with minimum trouble is that the operator should understand how it works. With the combine harvester that is easy; one can take it to pieces and put it together again. One can also conduct a post mortem on a cow, and, although it cannot be put together again, one can, with expert tuition, learn a good deal about how it works. The same thing can be done with a dog, but, for our purpose, we would learn nothing. Why? Because the characteristics which have, through the ages, made the dog such a valuable servant to mankind are mental, not physical. They cannot be seen. They can only be assessed and not always very accurately.

To assess them at all it is essential to give some thought to the evolution of the dog. In all classes of livestock breeding, man has achieved much of which he can be justly proud. And perhaps more than a little of which he should be thoroughly ashamed! Whether or no we approve of the changes he has made, one thing remains certain – only on rare occasions has he produced something which was not there to begin with or removed something which was. Cows which now produce phenomenal amounts of milk only do so because their wild ancestors produced milk to rear their young. Hens which lay an egg practically every day of their lives could not do so if the wild jungle fowl had not laid eggs to propagate the species. There must always be some foundation, and no one could, for instance, produce a breed of cow that laid eggs or hens which produced milk!

This principle applies much more to mental characteristics

than to physical, especially when dealing with instincts. Domestication, selective breeding and, to a lesser extent, training may weaken or strengthen an instinct out of all recognition, but they will neither put it there nor take it away. It can be diverted into other channels, just as a river can, but it cannot be reversed any more than a river can be made to flow uphill. I emphasize this because I have read statements by people who should know better that the sheepdog's instinct to herd is a *reversal* of the wild dog's instinct to hunt and kill. Nothing could be farther from the truth. The idea that a sheepdog herds because of an inborn kind-heartedness, and an 'almost human' desire to care for these supposedly stupid animals, gives rise to a great deal of trouble.

The instinct to herd is merely a divergence, in many cases a small one, of the wild dog's instinct to hunt. It should never be forgotten that, so far as we know, the domestic dog (be he Pekingese or St. Bernard) is descended from a mixture of rather horrible creatures like wolves, jackals and others. The dog thus springs from species whose survival depended on their ability to hunt and catch their own food. Had this not been so we should have no herding breeds today. I might add that if dog owners as a whole (including farmers) could be made to realize this, sheep and poultry worrying would become a very rare misfortune.

Returning to the wild dog, these almost invariably lived and hunted in packs. Very often the faster members would head off the quarry, turning it back to the slower ones coming up behind. Hounds, hunted in packs, will very often do the same thing, especially when the quarry is too large to be killed by one hound. The instinct to head off a quarry is the foundation on which the herding instinct has been built. By careful selective breeding man has strengthened and adapted this instinct until it is hardly recognizable. In much the same way he has produced the 2000-gallon cow from a wild animal that perhaps gave 150 to 200 gallons.

This instinct to herd is, perhaps, the most important of all to us. A dog with little or no such instinct *can* on occasion be trained to herd, but it is doubtful if it is ever worth the effort. As you will see later on, the willingness to herd forms the whole foundation of training. There is little point in laboriously building a foundation when it is easy to buy a dog with that foundation already there – a far stronger one than you could ever hope to build.

The whole advantage of a working dog lies in its ability to herd, or 'wear', as it is more commonly known. Someone once said to me, 'She works all right, but only when on the same side of the cattle as I am.' Practically any dog can be taught to rush backwards and forwards barking at the heels of some cattle. They do as little good and as much harm as the people one so often sees chasing livestock round the country-side with sticks. The dog's title of 'man's best friend', which he so much deserves, was won, not by being 'almost human', but by his ability to do things which man *cannot* do.

The 'strong eye' of the Border Collie also arises from the hunting instinct of the wild dog. The difference between the wolf or fox stalking its prey and the trial winner penning his sheep is not nearly so great as most people imagine. To try to demonstrate this I had a photograph taken of a Border Collie and a tame fox eyeing a tame rabbit (photograph 1). The greatest difference lies in the fact that, in the Collie, this instinct has been developed to such an extent that the dog is almost held back by its 'eye'. The fox reacts quite naturally, creeping near and then pouncing on its quarry (unless held back by a lead as in this case). But – and this is most important – at least 50 per cent of young Border Collies do exactly the same thing until they have been taught by the trainer that such behaviour is 'against the rules'. As a matter of fact, it is not *always* against the rules and the dog which simply lies and stares at a cow or stubborn old ewe is of little practical value to anyone – but we shall deal more fully with that later on. Sufficient for the present to remember that, far from being reversed, the hunting instinct has merely been diverted slightly and strengthened considerably.

To revert to the question of 'eye', this is a subject which is often misunderstood. First of all 'strong eye' is not essential in the working dog and I have known many excellent 'loose-eyed' dogs. It can be, and often is, a liability rather than an asset and, later on, I shall explain why. It is, in fact, a

comparatively recent innovation, having been developed to its
present-day strength only since sheepdog trials started, and
then solely in the type of sheepdog which proved most successful at the trials. This type, now generally known as the
Border Collie, has, by careful selective breeding, been developed into a distinct breed. Not fixed in type to anything like
the extent usually associated with a breed that has its own
carefully kept stud-book. The common factor lies in the style
of working, not in appearance.

Although 'strong eye' is a term usually associated with
sheepdogs, the instinct to set quarry has also been strengthened
by breeders of Pointers, Setters and, to a lesser extent,
Spaniels. I have been unable to find any evidence myself, but
some maintain that this 'eye' was introduced to the Border
Collie by crossing with the Gordon Setter. However, there is
evidence that working Collie blood was used in the evolution
of the Gordon Setter and that both Gordon and Irish Setter
blood were used in producing the show Collie. It is, therefore,
quite likely that a good deal of crossing did take place between
Setters and Collies and it is possible that that is the origin of
the instinct to set.

You may think there is a great difference between a gun dog
working with its nose and a sheepdog working with its eye. I
thought so too until that very famous trainer of sheepdogs,
W. J. Wallace, showed me something that has stuck in my
mind as the most uncanny thing I ever saw a dog do. He had a
blind dog working sheep and told me that he had another one
before which worked just as well. It is important to note that
the dog had gone blind after he had been trained, trained to a
very high standard, which enabled his handler to guide him
on his outrun to the sheep. Once he got on to his sheep, he
worked just like a dog with normal sight and I am sure that,
had they seen him, many of my readers would wish that their
dog could work as well.

At a distance experienced dog men would have commented
on the 'strong eye' this dog was showing, but his eyes were like
blue china and he could not see a thing. Obviously he was
using his nose to maintain contact with his sheep when downwind, and, I should say, his hearing helped him when upwind

of them. I need hardly add that the dog's sense of smell and hearing are so much more acute than ours that we really cannot understand how they work.

This experience set me thinking and the more I thought about it the more I wondered why it had never occurred to me that I had seen many sheepdogs use their nose as much as their eye. The very first sheepdog I ever trained would, if she winded a pheasant sitting in a clump of grass, creep up to it in exactly the same way as she would 'eye' a sheep. As the bird decided to take wing she would rush forward like a flash, springing in the air as it rose. It was a fifty-fifty chance she would bring it down but, as my father was tenant farmer on an estate where game was preserved, I could never brag about this accomplishment! These tactics were surely as near the wild dog as it is possible to find, but when working sheep this same bitch would go right through pheasants (I have seen them as thick on the fields as poultry) without so much as turning her head.

The points I want to emphasize are that Floss would set game which she *could not* see in exactly the same way as she would eye a sheep which she *could* see. Moreover, she would set a pheasant in very much the same way as a gun dog and I have seen many other Border Collies do the same. I have also seen a young Springer Spaniel eyeing hens just like a Border Collie. A dog gathering on rough ground may suddenly wind a sheep behind a rock or bush, stop dead in its tracks and 'eye' it when, in fact, it is out of sight. To illustrate this I put a duck in a basket, hid it in the middle of a whin bush and got my old bitch, Judy, to set it on the wind. Photograph 2 shows that she had adopted a pose much more characteristic of a gun dog than a sheepdog. To sum up, I have come to the conclusion that 'eye' is to a great extent a misnomer, although it is an expressive term which I shall be using frequently. It is really far more an attitude of approach than anything connected with the dog's eyes.

THE PACK INSTINCT

The hunting instinct is by no means the only instinct of the wild dog which man has put to his own use. Perhaps next in importance to us is the pack instinct. Most of the wild dogs hunted in packs. The result is that dogs will do things in the company of other dogs which they would never dream of doing on their own. That is unlikely to be of much help to you and is, in fact, the direct cause of a great deal of worrying of livestock. I mention it because it is so little understood, in spite of the fact that it is by no means peculiar to dogs. The pack or herd instinct is, in fact, as strong in humans as in any other animal and should therefore be easy to understand. Although dogs are never almost human, we are, in many ways, almost canine! Take as an example the rider who, in hot blood, sails over obstacles (and enjoys it) which neither he nor his horse would attempt in cold blood. When hounds are running and the field gets going, the herd instinct, in horse and rider alike, gives them a courage which neither possesses at any other time. The pack instinct has exactly the same effect on a dog, making it *an entirely different animal* in the company of other dogs from what it is on its own.

That aspect of the pack instinct is one which we have more often than not to guard against. Fortunately there is another aspect which is of the greatest value to us. No two animals are equal. In every herd or flock there is a boss or a leader with followers in order of merit, very often ending up with an unfortunate creature that is bullied by all and sundry. In no species is this more marked than in the dog, where we find a pack leader which does, in fact, govern and lead the whole pack. This he does, not by bullying the weaker members, but by putting in their place any who challenge his authority.

There are always in a pack one or two younger dogs which want to be leaders, but the majority are quite content to be followers, taking their orders from their superiors. Every now and again one of the young dogs will challenge the leader, who will, if he can, give the young dog a good hiding and put him back in his place. As time goes on, however, the leader will get older and one day a young dog will succeed in overthrowing him and gain the position of leader. If he has been leader for any length of time the old dog is unlikely to become a member of the pack. He will either be killed in battle, slink off to die or become a 'lone wolf'.

This is of the greatest importance in training. Just as the orphan lamb will attach itself to its human 'mother' and ignore the other members of its own species, so an animal, taken from its pack, can be persuaded to accept a human being as a substitute for a leader of its own species. As I have said already, the dog is not almost human but man can easily become almost canine and take on the role of pack leader. You need not be worried about that, as I am not going to suggest that you get down on all fours and bark like a dog! The late Konrad Most, one of the greatest German authorities on training, did, however, suggest tactics of that sort and there is no doubt that, to the dog, the human master takes the place of the canine pack leader.

To distinguish it from that aspect of the pack instinct which makes dogs want to do things together (usually getting into mischief), I like to refer to this particular aspect as the submissive instinct. It is because of this submissive instinct, and not because of its intelligence, as so many people imagine, that the dog is so much more easily trained than any other domestic animal. The cat, for instance, is no less intelligent than the dog, but, because of its independent nature and complete lack of submissive instinct, is extremely difficult to train. From the wild dog's instinct to obey a pack leader man has developed the submissive instinct, until today we have dogs which will instinctively obey a human master without any necessity for him to act like a dog. Providing, of course, that he is capable of making the dog understand what he wishes it to do.

There were, of course a minority of wild dogs born to be leaders, giving rise to the minority of domestic dogs which resent discipline. For some purposes, where courage and self-reliance are more important than obedience, these dogs are very valuable. In the working of sheep, however, implicit obedience is essential and, in the evolution of the herding breeds, it is unlikely that the occasional 'pack-leader type' would be bred from. It is, therefore, very rare to find a dog with a strong leading instinct in the breeds to which I shall be referring in the next chapter, but that does not rule out the possibility of your being unlucky enough to get one.

OTHER INSTINCTS

The instincts which we have been discussing are by no means the only ones handed down from the wild dog to his domestic descendant. They are, however, the instincts which have done most to make dogs so useful as workers of livestock. It must be remembered, too, that not all the instincts of the wild dog have proved useful to mankind. On the contrary there are some which man has had to breed out to the best of his ability.

Fear. One of these is the instinct of fear, common to nearly all wild animals. A natural fear of man (the enemy), of gunfire and a strong desire to keep well within range of bushes or cover of some sort. When taken from the nest at an early age many wild animals lose this fear. Our two badgers were more friendly with people (including strangers) than many dogs. They never attempted to run away on the several occasions that they escaped from their run.

This fact has enabled man to breed out this instinct to a great extent so that today we find dogs that are far more courageous than the average person. Many wild animals, however, no matter how early they are taken from the nest, never lose their natural suspicion or fear of mankind. People have taken fox cubs and brought them up like puppies. A few will allow strangers to handle them, but some resent being handled even by those who have reared them. Often, as they grow up, the urge to return to the wild becomes obvious and many succeed in doing so.

These, no doubt, were problems which confronted the prehistoric men who first tamed the wild dog. By domestication, and even more by selective breeding, these problems were to a great extent overcome. But breeders of all animals and plants must always fight against the tendency to revert to Nature. That is why, while anyone can breed scrub stock, the number of breeders like Booth and Bakewell, who will go down in history as great improvers of our breeds of livestock, are few in number. I doubt very much if, in the many aspects of livestock breeding, there is a stronger tendency to revert to Nature than in the mentality of the domestic dog. One has only to consider the mean, cowardly, furtive collection of thoroughly unlikeable creatures from which 'man's best friend' is descended to realize what that means.

I am not suggesting that a natural suspicion of strange people or objects is always a bad thing. On the contrary, the dog which keeps himself to himself is much to be preferred to one that rushes to welcome every Tom, Dick and Harry. But there is a vast difference between suspicion and fear, and the dog which bolts to hide in the darkest corner of the barn at the approach of every stranger is of no practical value to anyone.

Sex. The sex instinct is one which affects a working dog far more than is generally realized. You cannot expect a sex mad dog to keep his mind on working your stock. This tendency can, of course, be removed by castration, an operation against which there is less prejudice than there used to be.

A dog castrated *after it has reached maturity* is only changed sexually. His character remains the same in every other way, his guarding and working instincts are as strong as ever. I say that as one who has made a very careful study of the effects of castration in the dog and has probably owned and worked more 'geldings' than any other trainer. However, dogs castrated before maturity rarely mature. Failure to appreciate this fact is probably the cause of so much prejudice against the operation. I have recommended many people to have over-sexed dogs castrated with resultant peace of mind to both dog and owner. For many years all our dogs have been castrated with the exception of the few which we have wanted to breed from. Castration removes, to a greater or lesser degree, all the undesirable traits of the male dog – and there are quite a few!

Unlike the horse, the dog does not lose any of its desirable characteristics when it is castrated and no one has ever noticed that our dogs are all 'geldings'. In horses, where the vast majority are castrated with no prejudice against the operation, a stallion looks and behaves differently from a gelding. Most people prefer the latter but some, including myself, prefer the former and for many years now I have driven stallions in the show ring and driving competitions.

For many years now all our bitches have been spayed and we have not noticed any change in character. Many thousands of Guide Dog bitches have been spayed with no reports of adverse affects. Opinions differ as to the best age to operate. Some say that a bitch spayed before coming in season is liable to glandular trouble and obesity but that does not always happen. At the Guide Dogs for the Blind Association bitches are spayed after their first season.

The instinct to be clean. All animals born in a nest instinctively keep their living quarters clean. That is why pigs can usually be easily encouraged to use a dunging passage. As soon as they are old enough, cubs will leave the den to relieve themselves some distance away. Puppies will do the same – *if* they are given the opportunity!

If only people would remember that, much unnecessary cruelty could be avoided in the house-training of young puppies. A puppy reared in a kennel that is too small, or is not cleaned out regularly and frequently, has no alternative to being dirty and will become accustomed to living in filthy conditions. Such a puppy is likely to be difficult to house-train. One reared in a clean, roomy kennel, or which is able to go outside when it wants to, will get into the habit of going as far as possible from its bed to relieve itself. Such puppies are usually easy to house-train.

My dogs are reared as kennel dogs, but we seldom have a puppy over six months old that fouls his kennel and some are quite clean much younger than that. We never take young puppies to sleep indoors but can take practically any of our adults to live indoors at any time, and I can, and often do, take a dog with me to stay in an hotel or with friends without any fear of 'mistakes'. These dogs have no 'house-training' whatever. They have simply been fed regularly, exercised regularly, including a walk at 7 a.m. and 9 p.m. *every* day, and have been given the opportunity to develop their natural instinct to be clean.

If I sell a young dog that has taught himself to be clean he never presents any house-training problems to his new owner, on whom I impress the importance of sticking, as far as possible, to the dog's regular routine, at least until he has settled down.

DEVELOPMENT AND BALANCE OF INSTINCTS

The herding and the submissive instincts are among the most important factors which combine to produce a good working dog. Most important point of all is that the one should as nearly as possible balance the other. The herding instinct makes the dog want to herd, to use a common phrase, 'start to run'. The submissive instinct gives us what we call a biddable dog, which can be kept under control. The stronger the herding instinct, the more does it become necessary to have an equally strong submissive instinct to balance it. In breeding for trials, breeders have produced many dogs which have a quite abnormal herding instinct making them want to work literally anything that moves, from a cat to a double-decker bus. To keep such a dog under control, a strong submissive instinct is absolutely essential. On the other hand, a dog with a weak herding instinct may be far more difficult to keep under control if it has a proportionately weaker submissive instinct.

What you want, therefore, is a dog with as near as possible a perfect balance between the two. That, however, may not be very near, and you will have to make up the balance by using one or other of the equalizing factors at your disposal. The most important of these are plenty of work for the dog and your skill as a trainer. But I must emphasize the importance of getting a naturally biddable dog with a strong, but not abnormal, herding instinct.

The herding instinct is one which develops as the dog grows up and is not apparent in puppies in the nest. Sometimes it shows as soon as they start playing, and well-bred Border Collies will sometimes eye each other (just as their parents eye sheep) when as young as six weeks. Sometimes they will creep about after the hens when not much older but, if they do not, there is no need to worry. I once had a bitch that showed not the slightest inclination to 'run' until she was over a year old and I knew a dog that was eighteen months before he looked at a sheep. Both these animals turned out to be first-class workers.

It is, therefore, impossible to say at what age a dog will or should start to run. Quite often it starts suddenly and unexpectedly. The bitch mentioned above I bred myself, and she followed me around the sheep practically every day since she reached the age of three months. She would watch her mother working without taking the slightest interest. One day, without any warning and for no apparent reason, she left me, ran wide right round a field and gathered the sheep. In a week she was a useful worker, because I had given her her basic training and it was easy to make her understand what I wanted her to do.

It is of the utmost importance to remember that, when a young dog starts to run, he does so instinctively. When they see an untrained young dog 'wearing' a bunch of sheep many people say 'It's amazing how intelligent he is.' Intelligence has nothing to do with it. The young dog which suddenly decides to run can be compared to the young man who suddenly decides that a certain young lady is the most attractive he has ever seen! One does not have to be clever to do that sort of thing, and there are many who later wonder how on earth they could have been so stupid! The young dog herds, not so much because he wants to herd, as because he cannot help it any more than, when he was born, he could not help squirming around until he found where the milk came from.

At this stage I should, perhaps, clear up one of the commonest fallacies connected with working dogs: that it is necessary to have an old dog to 'teach' a young one. I have known young dogs learn from old ones all right but never anything I wanted them to learn! I do not know anyone who goes in seriously for the breeding and training of sheepdogs who relies on an old dog to start a young one.

Dogs do not copy or mimic members of their own or any other species. The fact that the dog often appears to do so is due entirely to the pack instinct and applies only to things which dogs do as a pack. The most important of these is, of course, to hunt. For this reason a young dog will often follow an older one. But he is just as likely to follow another young one, and two youngsters, neither of which shows much inclination to run on its own, will often run together. They will not, however, learn anything useful. All that happens is that, by being used, the herding instinct becomes stronger and, having been encouraged to start with another dog, the young one will probably then run on its own.

This I have quite often done when I have been anxious to get a youngster going but I am quite sure that, in most cases, the same goal would have been achieved by waiting until he started of his own accord. The risks involved in allowing a youngster to follow an older dog are considerable. These can often be avoided by using a little patience and waiting.

If he is allowed to run with his 'teacher' for any length of time the 'pupil' may refuse to run on his own. You may think that would put you back where you started, but it would really put you much farther back than that. What has happened is that, instead of you being accepted as leader, you have allowed the young dog to accept another dog. So long as the other dog is around you may have great difficulty breaking the bond.

By allowing two or more dogs to run together you are allowing them to go back to Nature. This encourages the true hunting instinct (which is why hounds are usually hunted in packs) and tends to weaken the herding instinct, so that you are really encouraging the youngster to go in and grip. This is a fault you are likely to have to cure in any case and there is no point in encouraging it to start with. In the well-bred dog the herding instinct should be strong enough to make the dog start without any outside encouragement. It is often helpful to have a reliable dog with you, not to teach the youngster, but to take control if he should be in trouble.

The submissive instinct develops much earlier than the herding instinct, and can be seen in very young puppies which will lick the hand they know in the hope of being petted. Unlike the herding instinct which, if left to develop on its own, will get stronger and stronger, this instinct tends to weaken as the dog grows up. The average puppy is only too delighted to follow any friendly human. Given the opportunity he will, as he grows up, become more and more submissive to the wishes of his master. Left to wander all over the countryside, however, doing exactly as he likes, the submissive instinct may practically die out, leaving a dog with little or no desire to please anyone except himself. Incredible as it may seem, no class of dog owner ruins such a high percentage of good dogs in this way as do farmers.

An instinct can be compared to a plant. With suitable treatment it will probably flourish and prove useful. Rough treatment in the seedling stage can easily kill it or so check it that very careful cultivation is necessary if it is to be kept alive. Some plants are, of course, so tough that short of digging them out it is almost impossible to kill them. Unless great care is taken such plants can get out of control and become weeds.

The abnormal herding instinct is difficult if not impossible to kill and, under certain circumstances, may well prove more of a nuisance than an asset. The normal herding instinct (the sort you want) usually goes through a seedling stage when it can easily be killed or badly damaged by trampling. A little care at this stage will result in a strong, healthy plant which can be trained and moulded to whatever shape you want. Generally speaking, instincts strengthen with usage, and if never used tend to die out.

This I have had the opportunity to study over a very long period and with a great many different dogs. Although I bought my first Corgi to work cattle, I did not for many years have any cattle to work. With one or two exceptions I have not found that this breed can compare with the Border Collie as a worker of other classes of stock. As most people know, the Corgi has no 'eye' but has a very strong heeling instinct, which means that instead of creeping about like most Border Collies do, the Corgi pup rushes about nipping everything by the heels. With several puppies being exercised with livestock around, it does not need much imagination to visualize the chaos that would result in allowing the working instinct to develop naturally. As I didn't want them to work, by far the easiest thing was to nip it in the bud, which is what we tried to do.

Of the several hundreds of this breed I have reared the vast majority, at some stage, showed that the working instinct was still very strong even when their pedigree showed no working blood for six or seven generations. Whenever a Corgi puppy showed signs of wanting to run we checked it and continued checking it until it gave up the idea.

Some showed no inclination to run at all except when encouraged by others and could soon be trained to ignore livestock. Others were so keen that, although they could be stopped, they never gave up the idea and would seize any opportunity to heel whatever happened to be handy. Those which concerned us most were the majority, which would work but which, having the herding instinct nipped in the bud, lost all interest and would run about amongst all sorts of other animals without paying any attention to them.

I have conducted several little experiments to see whether the herding instinct could be aroused again after it had been idle for some time. In some cases this was only too easy, in some it could be aroused only with difficulty and in others it was dead. Dogs which I know were keen to work as youngsters showed not the slightest interest. The seedling, once young, strong and healthy and with the right treatment ready to grow into a strong plant, had been trampled and killed. When not quite dead much more skill and care was necessary to rejuvenate the injured seedling than would have been necessary to keep it growing from the start.

The importance of all this lies in the fact that many dogs which will not work have had the herding instinct killed in exactly the same way as the young plant that is kicked out of the ground. Farmer A buys a well-bred pup and allows it to run around the place. It is not given anything to do, it is not even allowed to go indoors where it might find friends. It is left entirely on its own to follow instincts which will develop as it gets older. Sooner or later the herding instinct comes to life, bringing with it an urge which makes the pup want to herd. To herd what? Well, there are sometimes hens around but, if not, boys on bicycles make a good substitute or even cars do at least move! But Farmer A does not want a dog to work hens, and certainly not bicycles and cars, so he resorts to the standard cure for all ills and gives the pup a 'damn good hiding'.

Now Farmer A may well remember that in his youth he

resisted quite a lot of temptation to get into mischief, not because the Good Book said so, but because he knew that a 'damn good hiding' would be awaiting him if he didn't! Having been brought up on that principle myself I can fully appreciate how he feels, but there is a very big difference often overlooked. We were punished for doing something *we knew we should not do*. We may not have seen any real reason *why* we should not do it, but we had been told not to and were left in no doubt as to what would happen if we did it again!

In the case of the puppy, however, no attempt had been made to teach it anything. For doing what came naturally, and *not knowing it was doing wrong*, this canine child was very probably punished much more severely than would be necessary in the case of an adult dog being deliberately disobedient. This type of owner (I cannot call him a trainer) no doubt thinks that he has cured the pup of chasing cycles, which he may well have done. From the pup's point of view, however, he may simply have cured it of chasing. There is no real difference between chasing and herding, certainly not to a pup. The result is that, when Farmer A finally decides that it is time his young dog started working and takes it out amongst sheep, it will not go near them. Why? Because he has been 'cured' of that. The herding instinct is unlikely to be dead at this stage but, unless it is given plenty of encouragement, it very soon will be. The outcome is a dog that just follows his owner around, and an owner who, more than likely, blames the breeder for having 'sold him a pup'. Which is like grubbing out a crop of young kale plants and saying that you were sold dud seed. Depending on how strongly the plants are rooted and the thoroughness of the grubbing, they may or may not all be killed. What is certain is that you will not have done any good and that you will have only yourself to blame. I shall deal with what you should and should not do when we come to training.

2

Other Mental Characteristics

Temperament – Intelligence

TEMPERAMENT

At times closely associated with the instincts, but not one itself, is temperament, which is really a combination of a number of factors concerning a dog's mentality. It often combines with the submissive instinct to give us 'hard' and 'soft' dogs. The independent pack-leader type is often bold and courageous but hard and difficult to train. The very submissive dog, although easily trained, is often soft, and lacks 'guts' when put to the test. Likewise, temperament and the instinct of fear combine to give us dogs which tend to be shy of strangers and, unfortunately, dogs which are terrified of everybody.

Temperament in the working dog is much more important than is generally realized. To the hill shepherd it does not matter so much as he is usually on his own. So long as the dog does not bolt at the sight of a stranger it may be a first-class worker. Many good hill dogs are, in fact, shy partly because of temperament and partly because they never see strangers. On the general farm, conditions are very different. As the peace and quiet of the countryside becomes more and more a thing of the past so does it become more and more important to get a dog with a good temperament. This, as I know to my cost, is a characteristic which is liable to change as the pup grows up.

To keep up a team of demonstration dogs, as we did, required constant replacements. Of the puppies which we selected at eight weeks (taking into account the pup *and* the breeding) as being suitable for our job, I doubt if 50 per cent ever appeared in public. By the time they were six months to a year old we had reluctantly decided to discard the other 50 per cent. Although this was in some cases due to otherwise good dogs not growing up as we should have liked them to look, in most cases it was due to their not having the sort of temperament to put up with the people, the noise and the hustle bustle they would have to endure in our job.

You will, no doubt, be more easily pleased, but I should like to emphasize that the pup, which at eight weeks was as bold as brass and which is bred from equally bold parents, may end up shy and timid. Upbringing, to which I shall refer later, undoubtedly plays a big part, but not as much as some people think. I have known a shy puppy improve in temperament but only rarely, and I cannot over-emphasize the desirability of getting a bold puppy bred from equally bold stock.

There is another very important aspect concerning temperament, one all too often overlooked. A dog that will work well for one person may not work at all for another simply because they are temperamentally unsuited to each other. Some people, who can train a hard dog, can make nothing at all of a soft one and vice versa. A dog should, therefore, have the sort of temperament to suit it, not only for the work it will have to do, but also to the sort of person who is going to work it.

As people vary in temperament just as much as dogs, it is impossible for me to say what sort of dog I think should suit you, but the following points are worth remembering. A 'hard' dog is strong-willed and the stronger the will-power of the dog, the stronger is the will-power necessary to control it. Many of the problems with which I have to deal arise through the dog having more will-power than the owner. No better servant than the dog has ever been found, but it can be, and often is, a very bad master. This problem does not, however, affect readers of this book as much as those of the other books I have written. The weak-willed sentimentalist does not, as a rule, choose farming as a career. If he does he rarely has the willpower to overcome the many snags. Although farmers 'as a breed' are notoriously bad trainers, this is not usually due to their allowing the dog to become the boss.

The 'soft' dog is much more easily trained. The trainer requires much less will-power. but he requires much more patience. Anyone who loses his temper with any animal should keep right away from them. If you are that sort I suggest that, for the sake of your animals and yourself, you stick to your tractors and leave the handling of stock to someone with a controllable temper. At the same time, there are many people who, without losing their tempers, simply have not the patience to coax and wheedle a soft dog. Moreover, in these times, few people in any walk of life have the time to use their inexhaustible patience. But some people do like a more sensitive type of dog even if he is a bit soft. In the right hands this type will respond to the slightest whisper and is quite fascinating to watch or to work. In the wrong hands it won't do anything!

In considering the type of temperament to suit you, note must also be taken of the type of work you want the dog to do. A hard dog thrives on plenty of work and, as I shall explain later, the less work you have for it the more training you will have to do. A soft dog is usually quite easy to control even when it has spells of idleness. Here I should like to repeat some advice given by the well-known Blackface sheep breeder, Captain James Craig of Inergeldie, Comrie, to a party of young farmers, of which I was a member. 'Never go to extremes,' he said. Although at the time referring to the changing fashions in show animals (colour of the 'black' face in this instance) I have found it a very good maxim in *everything* connected with animals.

In choosing a working dog, therefore, do not pick one that is too hard or too soft but try to get something between the two extremes. For your purpose one a bit hard is always likely to be better than one on the soft side. In an emergency a hard dog will always do *something* whereas the soft dog may well turn and run away. If you are loading cattle into a truck and a heifer squeezes past the ramp the hard dog will be there to try to stop her. Even if he cannot turn her he will stick to her head and slow her up until someone catches up to help. With all the shouting and carry-on associated with the loading of cattle on the average farm the soft dog will probably have retired to a a safe distance. Even if he decided to try to stop the heifer in the first place, by the time one or two bright lads with sticks catch up with her he will almost certain retire from the fray just when he is most needed.

That sort of thing always happens when the truck has arrived late or you have been held up by a cow calving or a breakdown in the harvest field – never at a time when you are in the least inclined to humour a temperamental dog! To me, and I feel to the great majority of my readers, a dog that will 'have a go', even if he does the wrong thing, is never half so aggravating as one that gets flummoxed and just looks on or disappears completely.

INTELLIGENCE

Some people may be wondering why I have not yet dealt with intelligence, considered by many to be the most important quality in any dog and by some the only essential. I have two reasons. Firstly, intelligence is in no way connected with any of the characteristics we have been discussing. Many people are unable, sometimes unwilling, to differentiate between intelligence and instinct which are quite different things. Stupid dogs are just as keen to run as intelligent dogs. Likewise stupid *and* clever dogs may have good *or* bad temperaments. Many very intelligent dogs are regarded as stupid by people who are themselves too stupid to realize that the dog simply lacks confidence. Secondly, I doubt if any of the dog's admirable qualities is more over-rated than intelligence. Nearly all 'problem' dogs are intelligent, often so intelligent that they make rings round their stupid owners. Of two young dogs equally keen to run, one intelligent, the other not, the intelligent one will learn more quickly than the other. An intelligent dog that will not run, however, is of no more use than a cat, whereas the keen dog, even if he is a bit slow in the uptake, may end up a useful worker.

Intelligence is not one of the characteristics which man has improved by domestication. Most country people know how clever the fox is, and naturalists tell us that the intelligence of this animal is much lower than that of the wolf. If intelligence

Showing 'eye'
1 Border Collie controlled by training; tame fox controlled
by force

2 Judy 'eyes' something she cannot see

3 Corgis *do* work cattle

4 Have nothing to do with the pup at the door, be very doubtful about the one on the left and pick from those which come to you

5 and 6 A good dog should walk quietly behind cattle
so long as they are moving, but should be ready to head
any that break away

7 Severe correction for wilful disobedience

were the only essential in a working dog it would be just as
easy to teach a fox to work sheep as to teach a collie. That it
certainly is not proves that intelligence is only *one* of many
characteristics which go to make up the complex mentality of
the dog.

All the same, do not get the idea that intelligence is of no
importance. Remember, however, that just as the wild dog
used its intelligence to help it capture its prey *and* to evade
capture by its enemies, so can the domestic dog use it for *two*
purposes. It can use it, and we hope will use it, to learn what
we are trying to teach it. Unfortunately, it can, and very often
does, use it to find ways and means of *not* doing what we want
it to do. Intelligence does not, of necessity, make a dog easily
trained and can on occasion make it quite untrainable.

Provided a dog has a reasonably strong submissive instinct,
making it want to please a master, intelligence can be a
tremendous asset. Even if this instinct is not as strong as it
might be, an intelligent dog will soon learn that, even if he
does not particularly want to please his master, it is to his *own*
advantage to do so!

There are, of course, different types of intelligence in dogs
as in humans. There is the super intelligent 'highbrow'; sensitive, often touchy but, in the right hands, quite brilliant. Then
there is the level-headed 'lowbrow'; not so brilliant but with
much more common sense. I hardly need emphasize that
conditions on the general farm are rarely congenial to
highbrows on two legs or four.

Jock Davidson, who initiated me into the art of handling
both dogs and sheep, used to tell of a relative of his, a retired
seafaring man, who had taken up farming. He had a quite
exceptional working dog but when it made a mistake the
owner would shout 'Go home, you fool, go home.' 'But,' said
Jock, 'the dug had mair sense!' Without wishing to insinuate
that any of my readers are of that type, I do think that that is
the sort of dog which will suit the majority.

Intelligence and common sense are more important in a
farm dog than in a hill or trial dog. In the hill dog the instinct
to go out and gather is the first essential. The trial dog is con-
tinuously under the control of the handler who gives constant

instructions. The farm dog, however, has very often to think for itself, it is expected to do many jobs that the hill or trial dog never attempts and, perhaps most important of all, is often expected to work for any Tom, Dick or Harry. That many dogs have enough intelligence and the right sort of temperament to adapt themselves to such conditions proves just how adaptable the dog can be. That many do not is in no way surprising, and trouble may be experienced in finding one that will.

Intelligence, and initiative which often goes with it, show in the puppy at an earlier age than any of the other characteristics. Although impossible to tell whether it will run or not and difficult to tell what sort of temperament it will have, the most intelligent pup in a litter is sometimes obvious at between three and four weeks old. It is the first to crawl out of the nest, quickest to learn the voice of whoever feeds it, finds ways of getting out of an enclosure that keeps the others in and gets itself into all sorts of mischief. From the study of many litters I have found that a puppy of that sort nearly always retains its mental lead over the others. This does not mean that it is bound to end up the best dog. Often the pup which is a bit slower to learn is steadier and more reliable. He may be lazy but I have found that lazy dogs, like lazy horses, are rarely stupid. For many purposes they make far more useful servants than those that can never keep still.

3

Balance of Mental Characteristics

SO FAR we have been discussing the mental characteristics with which the dog is born, but dogs acquire characteristics, good and bad, as they grow up. This, to a great extent, depends on *how* they are brought up. In other words it is up to you to see that the good plants develop and the weeds do not become strong enough to smother them.

Buying a dog is, in many ways, like taking a farm. The first thing is to decide how much money you can spend. But money cannot buy everything and, even if you have the money, you may have to be satisfied with something far short of your ideal. Farms, of course, are not judged by the soil, the situation or the layout alone but by a combination of them all and many other factors. What might suit one man would not suit another. Most people adapt their policy to suit the farm as much as they adapt the farm to suit their policy. It is all a matter of give and take, of balancing the good points against the bad, making every use of the good and trying to improve or eradicate the bad.

All that can be applied to a dog and, if more people went to a little more trouble to find the right dog to start with, there would be fewer disappointments. Even if you do go to the trouble of finding a dog with the various characteristics as nearly as possible balancing each other, it is still up to you to make up any deficiency and to see that the desirable traits are encouraged to develop and not vice versa as so often happens. Many men with limited capital have made a great success of very poor farms and many men with plenty of money have made an incredible mess of first-class ones. In the same way, one man can buy a mediocre dog at a small price and turn it into a useful animal, while another can pay a big price for a first-class animal and completely ruin it! Whether you are a good trainer or a bad, however, if you start off with a good dog you will almost certainly end up with a better dog than if you start off with a bad one.

4

The Right Dog

WHEN asked what is the first essential in training, my reply is always the same – getting the right dog to train for the particular purpose in mind. The first thing to decide, therefore, is what you want the dog to do. Many dogs could (and would if given half a chance) do far more than their owners ever imagine. By some trick of fate, dogs with the greatest potential abilities often seem to be landed with owners who cannot or will not train them. They are like clever children who are not allowed, or not made, to go to school. Conversely, many people spend much time, often money too, on dogs which simply haven't 'got what it takes'.

As in many other things, perfection often necessitates specialization and, for those who want perfection, the obvious thing is to keep several dogs, each expert at his own job. To the average farmer, however, that is quite impractical and my object at this stage is to help you find a dog which, with as little training as possible, will have a go at any job that happens to crop up. He may not be quite such a polished performer as the specialist dog, but then I am not, in any case, writing for the person who can apply the polish.

What I have in mind is a dog that will work sheep, cattle, poultry or any other type of stock which happens to be on the place. The various illustrations with my old bitch Judy will, I hope, help to convince those who doubt that the same dog will do so many different jobs. Apart from the jobs shown, she is a keen retriever and would, when she was younger, attack a man on command. She has had many film parts and, because of her reliability at barking on command is heard, but not seen, in a number of others.

If you have only one class of stock, obviously it is not essential to obtain a dog that will work all stock. Although many dogs (far more than most people realize) will work practically anything, few will work all equally well. You want to make sure, therefore, that the dog will work the stock you keep. To work cattle, especially heifers or bullocks, it must heel and nose when necessary. But it does not follow that a dog which will grip cattle will necessarily grip sheep. As I shall explain later, a dog should never grip anything until told to do so. The dog which cannot be stopped gripping sheep rarely makes a good cattle dog.

I was brought up on an arable farm in East Fife where my father fattened about 100 Irish bullocks every winter. He was quite convinced that a dog which would work sheep could not possibly work cattle. I have seen a dog bite the ear clean off a bullock! That is not really the sort of dog I advise anyone to keep! At the same time remember that, in training a dog, you can only work on what is there. If a dog has not got the 'guts' to face a bunch of bullocks no amount of training will make him do so.

For myself I always like a well-bred, 'strong-eyed' dog. People who have worked such dogs rarely get much pleasure from a common-bred one. It is like riding a hairy-heeled cob after a blood horse. But don't forget that, in some countries, the latter is much more likely to break his own and his rider's neck than the former. Also the blood horse is of little value to anyone who cannot ride him, and there are many people who will see far more of the hunt, with much less effort, from the back of the cob.

As the only British breed of sheepdog with 'strong eye' is the Border Collie, I shall be dealing with the subject more fully in the next chapter but should like to repeat here 'Never go to extremes'. Deciding what you want the dog to do is usually far easier than finding the dog to do it. If the only essential is a dog to do a job of work, then by far the best thing is to buy a trained dog. Good dogs, like good farms, are often hard to find and, when they are found, they usually cost a lot of money. Even then, from a purely economic point of view, they are usually the best investment.

In sheep-farming districts it is sometimes possible, though increasingly difficult, to find 'hill-worn' dogs at reasonable prices. These are dogs, usually five to seven years old, getting a bit slow for high ground but very often ideal for work on the general farm. If a shepherd keeps a dog to that age you can be pretty certain that it is a good one. Being a seasoned worker that really knows its job, it is by far the best for a handler who does not know his! Even at the same price as a young trained dog these older dogs are often the better bargain, especially for a beginner.

Many hill dogs never see cattle but this does not mean they will not work them. It does mean, however, that, if you want a dog to work cattle, you must make sure that the dog you choose will do so. Another snag, more imaginary than real, is that some hill dogs run so wide that, on a lowland farm, they may miss your sheep altogether and gather your neighbours! A dog very soon learns his own ground and, in fact, his own stock, and, although difficult to get a close-run dog to go out, it is usually easy to get a wide-run dog to come in. With a little patience in helping the dog to adapt himself to the new conditions, this never proves to be a very big problem.

Unfortunately there are not enough of this type of dog to supply the demand today (I can remember when they could be bought easily for £1) and you may have to look for something else. There are quite a number of people who make a business of buying young dogs, breaking them and selling them as trained dogs. Some start them working and turn them over as quickly as possible, so that, although the dog is trained, he is not experienced. He is like the young man out of college who still has most of his learning to do.

There are other breakers who keep their dogs much longer, using them in everyday practical work. When they have a youngster ready to take over they will sell the trained dog which has perhaps had a year or more practical experience. Such dogs command big prices and many are now sold at auction for ordinary farm work at prices comparable to good dairy cows. Very often they are worth it – that is to say if you have enough money to buy one!

If you decide to buy a trained dog I strongly advise going to see the dog on its own ground, not only to see how it works but to learn how to get it to work. It is a common practice to sell working dogs on a week's trial, but I think this unfair to both buyer and seller and grossly unfair to the dog. A highly sensitive and intelligent animal is suddenly taken from its native Scottish hill and eventually reaches an owner who speaks a language it has never heard before.

No one would expect a young shepherd from the same hill to settle down with a new master and show his skill with a flock of Down sheep *in a week*. The fact that many dogs do settle, some literally teaching their new masters how to work sheep, merely proves the dog's superiority over man in controlling livestock. Even so, it is not surprising that many excellent dogs are returned as useless, giving no satisfaction to either buyer or seller and much unnecessary suffering to the dog.

A fortnight's trial is not so bad, as it gives the dog time to settle down, but even then I strongly advise seeing him working at home first. If he does not prove suitable you will then know that it is not the dog to blame although you are not necessarily to blame either. On more than one occasion I have seen a dog working, liked him, thought he would suit me, but, on getting him home, have found that he did not take to my job at all. The best thing, therefore, is to see the dog working, learn as much as you can about how to work him, and then have him for at least a fortnight's trial.

If you do not have enough money, you may have to buy something untrained and make the best of it. Of course that is what I hope most of my readers will do, as otherwise there would be little point in writing this book. I have been careful to point out that a trained dog is the best bargain from a *purely economic* point of view, but, if we all looked at life from that angle, many of us would be working in factories or perhaps as navvies. The economists tell us that, by doing so, we would be greater assets to ourselves and the country. That few take this advice shows that, even today, there are greater satisfactions in life than just making money. Some of the greatest satisfaction I have had has been in successfully training dogs and horses which have beaten other people. But I rarely made money out of them – rather the reverse. I doubt very much if anyone who values his time at present-day rates has ever made money out of training young dogs. If, therefore, your main object in life is to make money, buy a trained dog.

For those who want to train a dog from scratch there is no shortage of raw material. Compared with the prices of other animals, especially other classes of dogs, this raw material can be obtained at a ridiculously low cost. Usually the best bargains are to be found amongst young dogs ready for training at, say, six to twelve months old. There are always people buying puppies with the good intention that they are going to train them. Because the pup wants to work it becomes a nuisance. By that time the owner has often lost the inclination to train it, makes the excuse that he hasn't time, and offers it for sale, sometimes at less than he paid for it.

By then you can tell, to some extent, how keen the pup is likely to be, how much 'eye' it will have and what sort of temperament it has. One with any tendency to shyness or to slipping round the back of a stranger and 'legging' him should be avoided at all costs. It is also well worth testing for gun sureness. There is no need to fire a twelve-bore over its head, but do make sure that, if the gamekeeper fires a shot in the wood, the dog is not going to bolt for home.

The 'trial men', who invariably rear dozens of pups for every one that ever sees a trial ground, often have young dogs to sell just 'starting to run'. These dogs are obviously not going to be up to trial standard but they may well be as good or even better for your job than their more flashy brothers and sisters. You will not buy a young dog from these men for the price of a pup but it is likely to be worth the extra money. It has probably had a certain amount of initial training and most certainly will not have been allowed to develop bad habits.

In buying a young dog make quite certain *why* it is being sold as, although many genuine youngsters are on the market, there are invariably a number which have proved useless to their owners and are likely to prove just as useless to you. The 'good intentions' type of owner often allows a pup, perhaps a good pup, to develop bad habits. Faults can usually be eradicated if met as they appear but, if allowed to establish themselves, they will prove difficult, perhaps impossible, to get rid of. There are many people who, having allowed and even encouraged a pup to develop bad habits, and eventually deciding that they can do nothing with it, look around for an unsuspecting novice to whom they can sell it as 'keen to run'.

On seeing a dog for the first time it is often difficult to judge whether the mistakes he makes are due to inexperience or whether they are established habits. The inexperienced young dog is usually somewhat hesitant and gives the impression of not being quite sure of what he is trying to do. The dog which runs in a determined way and persists in doing the wrong things has very likely got into the habit of doing so. The wild young dog which pays no attention whatsoever to its owner, but which shows the instinctive abilities we are looking for, is infinitely to be preferred to the much more obedient animal which has been started the wrong way. You can, for instance, instil obedience into a disobedient dog far more easily than you can instil confidence into a stick-shy one. When we come to training I shall be referring to the usual faults that appear in young dogs. As I shall also be dealing with the difficulty or otherwise of correcting them, I need not do more at this stage than warn you against faults that have become habits.

We have not discussed the age at which most people do, in fact, buy a dog – as a young puppy. There are two main reasons for that – firstly it is the age at which most breeders want to sell them and secondly many people think, quite rightly, that there is no risk of its having been 'messed about' before they get it. In spite of this, a young puppy is undoubtedly much more of a pig-in-a-poke than a young dog. On the other hand, to those who like dogs, a great deal of satisfaction is to be derived from choosing a young puppy from a litter, rearing it and training it from scratch.

Selecting a pup that is likely to turn out a good worker is rather like choosing a heifer calf that is likely to turn out a good milker. Its breeding is by far the most important factor. That does not simply mean a long pedigree – it means finding a pup in whose ancestry are the sort of dogs you would like that pup to grow into. Having found a litter with the right sort of breeding you may, if there are several for sale, be able to pick your individual pup.

Most breeders of working Collies try to sell the pups as soon after six weeks as possible. Although they should be weaned from their dam by then I consider it too young for a puppy to have to go and live on its own in a strange place. I always keep mine till they are seven or eight weeks, and a week makes a tremendous difference to a pup of that age. We will assume that the litter is from six to eight weeks old, an age at which it is usually impossible to tell whether a pup will work or not. I say 'usually', because I have occasionally seen eight-week-old pups showing 'eye', but they often turn out far too 'strong-eyed' for general use so that does not help us much.

Although the only guide as to whether or no a pup will have any herding instinct lies in its breeding, one can form some idea of intelligence and temperament from the puppy itself. Many shepherds I know, when they go to select a pup, simply pick up the one which reaches them first. That method is not nearly so much of a lucky dip as it sounds. The bright puppy is nearly always first off the mark and the bold one will go straight up to anyone, so by adopting this method you are likely to pick the brightest, boldest pup, which is just what you want.

Even if you cannot be sure that the bold pup will remain bold, you can at least avoid the shy one which has every chance of remaining shy. If I clap my hands in front of a litter of puppies about half of them usually run away. I am not saying that those pups will be gun-shy but I should always choose one of those which remain. Conditions under which puppies are reared should, of course, be taken into consideration. I have known puppies, born and left to themselves in a dark corner of a barn, which at six weeks were literally as wild as fox cubs. I have also seen lots of pups reared in the kitchen and played with by children which, at the same age, had no fear of humans at all.

It is after that age, however, that the individual temperament develops, and one could be wrong in assuming that the bold puppy in the kitchen would finish up a bolder dog than the shy one hiding in the barn. Comparison can be a great help. If one puppy has obviously a better temperament than others of the same litter then, all things being equal, that is the one to have. Even more so, if a litter comes running out and one, on seeing a stranger, bolts back to its kennel or sits in the corner looking coy like the one in Photograph 4 – *leave it there.*

The majority of shepherds like to have a young dog starting to run soon after the lambs have been sold. The ewes are then in good heart and it will not do them much harm if they do get chivvied a bit. In fact, it may prevent their going to the ram with too much flesh. By the time the tupping season is over and the ewes begin to show in lamb the young dog should be well under control. By lambing time he should be quite an asset. This is a point worth bearing in mind when buying a pup. But remember that, if your pup wants to run at six months and you are not ready to start him, you can wait until he is nine months without doing any harm. If, however, you plan to start him at six months and he does not run for another three there is nothing you can do about it except wait. The vast majority of dogs are running by the time they are nine or ten months old. If you plan to start your pup at that age and it does not run until it is a year old you will have been unlucky. If, on the other hand, you plan to start him at six months, and he will not run for another three or four, you will simply have been foolish.

A young dog should be given some work every day if possible. This is especially so when he is just starting to run. If a pup starts to run just as harvest is starting it will probably be much better to keep him off work until it is over, when you may find it easier to give him a turn every day. These points are all worth remembering in deciding when to buy a pup.

The idea that winter pups do not do as well as spring pups is not quite true. Provided they have a dry house and are well fed, with cod-liver oil added, they should do just as well.

To what breeds should we look in the hope of finding the sort of pup we want? There must be somewhere in the region of fifty breeds of sheepdogs used in various parts of the world today, but it should be remembered that different people have different ideas of work. In this country there are hundreds of dogs regarded by their owners as working dogs whose only claim to working ability is to bark at the heels of some cattle, a task that the average terrier or mongrel would be delighted to take on. In some countries abroad, sheepdogs are kept as protectors of the flock and will instinctively stay with the sheep. They have little herding instinct, however, and are rarely used to work as we would want.

In view of the fact that Great Britain has produced most of the world's greatest breeds of farm stock it is not surprising that it has also produced some of the best dogs to work that stock. I therefore intend to confine most of my observations to our own native breeds. That does not mean that if you have an individual of a foreign breed, it cannot be taught to work. German Shepherd Dogs, in particular, have proved excellent workers of various types of stock, and sheepdog trials are held regularly for this breed in Germany. These trials, however, are very different from those run by the International Sheepdog Society. Although the Germans have taught us a lot about training Police and Service dogs, I do not think these extremely clever trainers can teach us much about working sheepdogs.

Of the breeds and types native to, and commonly found working in, this country by far the best known today is the Border Collie. At one time known as the Trial Bred Collie, Creeper, or Working Sheepdog, this is the only British sheepdog which has its own stud-book (kept by the International Sheepdog Society). The first stud-book was published in 1910 but that could more accurately be described as a collection of names than a record of pedigrees. The first International Sheepdog Trial was held in 1922 and it is really since then that the Border Collie started its great increase in popularity. It is now to be found in every corner of the British Isles and in every country in the world where sheep are kept.

It is not really correct to describe this as a type of Collie as type varies tremendously and it is impossible to distinguish between a true Border Collie and several other types by simply looking at them. The real difference lies in the style of working, aptly described by the term 'creeper'. These dogs work with 'eye' and style found in no other sheepdog except the Australian Kelpie which was bred from Scottish stock and is really a strain of Border Collie. One finds a wide diversity of type from the big, rangy, bare-skinned dog which, to the uninitiated, looks more like a lurcher than a collie, to the pretty little shaggy type not much bigger than a Cocker Spaniel. Ears vary from pricked like a German Shepherd Dog to dropped as low as a Labrador. And they are not all black and white as many people imagine. That is certainly the predominating colour but I have seen pedigree Border Collies black and tan, sable, red, chocolate, chocolate and tan, blue merle, slate blue and blue-grey all with or without white markings.

A few dogs registered in the stud-book are 'Beardies' and dogs of this type have occasionally been up to 'International standard'. These are throw-backs and not true Bearded Collies. Apart from coat texture they are in every respect typical Border Collies. I had a pup Ben, very well bred and registered in the stud-book, which was nearly all white, as well as having a Beardy coat, although there are none of this type for several generations in his pedigree. This pup was only one of many examples which prove how impossible it is to describe what a well-bred dog looks like. Unless, of course, he is working, when his good breeding becomes obvious to anyone with experience.

A common but mistaken idea is that the Border Collie is Welsh and that the name is derived from the borders of England and Wales. This is probably due to the fact that the very first sheepdog trial was held in Wales in 1876, since when Welshmen have shown such a tremendous enthusiasm for trials (and skill in training dogs for them) that the Border Collie has almost ousted the native Welsh variety. That does not, however, alter the fact that these dogs originated in the borders of England and Scotland (the first trial in Wales was won by a Scotsman). To describe them as Welsh is as far off the mark as if I were to describe as Scottish or English the many Welsh Corgis I have bred in Scotland and England.

The twentieth century has probably seen the most marked advance in the breeding and working of sheepdogs which has ever taken place since man first used the dogs to help him herd his flocks. There is no doubt that much of this advance is due to the cult of sheepdog trials and to the work of the International Sheepdog Society. As this book is not intended for

the trial enthusiast, and as I can never see the point of filling
books with statistics copied from other publications, I shall
not go into details of the trials here. For those who are interested
– and I hope many readers will be – full particulars can be had
from the Secretary of the International Sheepdog Society.

Since this book was revised in 1975 millions of people who
had never seen a sheepdog trial (nor are likely to see one) have
sat enthralled at the highly successful TV series, 'One Man and
his Dog'. This has helped to create what could be described as
the cult of the Border Collie with many people wanting to own
one who have absolutely nothing for it to do.

Previous to that, the Border Collie had gradually become
the most successful of all breeds in Obedience Competitions
at dog shows. Not because of its superior intelligence, as some
people would have us believe, but because of its greater
trainability. As I explained earlier, there is a vast difference
between the two. I have serious doubts about the intelligence
of those people who teach dogs to walk along bumping up
against their owners' left legs and gazing up into their faces
like demented idiots. It certainly does not point to any
intelligence in the dogs. But that is what one has to do to
satisfy present-day obedience judges in the UK.

Another thing that has happened since my last revision is
that the Border Collie has been recognized as a pure breed by
the Kennel Club. Which means that it is now being exhibited
in beauty classes as well as obedience classes at dog shows.
And for the first time it is being bred for looks rather than
working ability. Not surprisingly there was strong opposition
to this move from many working sheepdog men who were
(probably still are) convinced that the breed would be ruined
as a worker. I was very much against the Kennel Club recognition, but for another reason. Popularity as a show dog
usually means popularity as a pet, for which purpose the Border
Collie is one of the most unsuitable of all breeds.

There is no reason why being accepted as a show dog
should ruin a breed for working. What usually happens is that
show strains and working strains develop alongside one
another and it is up to buyers to decide what they want the
dog for. It is interesting to note that in Australia the Border

Collie has been popular as a show breed since the 1950s. Even more surprising is that the native Australian Cattle Dog has been shown since 1904 and the Kelpie since 1908. The two latter breeds are unsurpassed at their own job and there is no scarcity of good working Border Collies either.

Many people have benefited from the cult of the Border Collie. Some 6500 pups are registered annually with the International Sheepdog Society, which is now in a stronger financial position than ever before. This book sells as well as it did when first published over twenty years ago. Several other books have also been published – indicating the wide interest in sheepdogs and their training. There is a ready market for registered pups at much higher prices than of old.

It is also possible to sell Border Collies which have proved useless as workers and which would previously have been put down. Provided they are up to Kennel Club standard of points they will be in demand as prospective show dogs. Others can be sold as pets or for training for obedience competitions. There is nothing wrong with that, I have sold dogs for those purposes myself. Indeed, the dog that refuses to run is likely to be as great an asset, and much less of a liability as a pet, than one that is overkeen to work.

What worries me is that the International Sheepdog Society makes no attempt to divide the registration of pups bred from working stock and those bred from stock which has never worked. The stud-book of the I.S.D.S. is a 'closed book', which means that no pup can be registered unless it has registered parents. For instance, I might have an absolutely outstanding, unregistered bitch, which I know is bred from a line of good workers – registered or unregistered. I could have this bitch mated to the best registered working dog in the country, but the pups would not be eligible for registration, nor would their offspring be eligible. On the other hand I could have a registered bitch that had proved quite useless as a worker. She could be mated to a registered dog which was winning in obedience or the show ring, but had never seen a sheep. Nor had his parents, grandparents or possibly great grandparents proved themselves as workers. But the pups can be registered without any trouble.

The majority of Border Collies being shown or competing in obedience are registered with the I.S.D.S., but some have no working blood in their pedigrees for four or five generations. The number of these dogs is increasing quite rapidly and, as it does, so the number of registered pups bred from non-working strains increases even more rapidly. I tried to persuade the I.S.D.S. to keep separate registers for pups bred from working and non-working stock, but to no avail. All I can say to you is that if you decide to buy a registered pup make sure it is from a working strain. Being bred from trial stock such a pup can be expected to have some 'eye', although one can never say how much. Nearly all 'strong-eyed' dogs run wide and perhaps at this stage I should try to throw some light on this subject so often misunderstood.

A well-bred pup, when he starts to run, should (I do not say he invariably will) cast out in a complete half-circle either to right or left (Figure 1). Leaving the handler at A he will end up at B, a point slightly more than half-way round the circle. If he is a really good one he will do it without any training or instruction whatsoever – he will do it instinctively because he has been bred to do it. Having reached point B he should stop and then approach his sheep in a straight line. If the sheep move to the handler's left the dog moves to his right and vice versa so that no matter how the sheep move they are always between the handler and the dog.

The advantages of this should be obvious. The wide-run dog does not disturb his sheep at all until he comes on to them from point B. If he comes on quietly and keeps well back they will come on quietly. If he moves quietly to either side *as the sheep move out of line* he will steady them and keep them in a bunch straight in front of him. The close-run dog, on the other hand, will run straight from the handler to the sheep. The latter, on seeing the dog rushing straight at them, will obviously run away with the dog in pursuit. If the dog passes on the left, the sheep will wheel to the right with the dog close in on their flank. By running them in a circle the dog may eventually get the sheep to you but no one will ever be able to estimate the damage that such a dog can cause on a flock of in-lamb ewes.

Some common-bred dogs make matters worse by barking as they run, thereby not only chasing the sheep all the faster but scattering all your neighbour's sheep as well! Of course, a dog should never be allowed to do that sort of thing, but it is far easier to buy one that won't than try to train one that will. 'Try' is usually the operative word! Normally it is easy to get a wide-run dog to come in but extremely difficult to get a close-run dog to go out. This is especially so in the confined areas in which most of my readers will have to work their dogs, and my advice is to select one that runs wide and if necessary teach him to come in.

Another advantage of the 'strong-eyed', wide-run dog is that he is much less likely to bring half the flock and leave the rest. Supposing the sheep are scattered all over a field (Figure 2) and the dog is sent out to the left from A. The wide-run dog will cast round the first bunch of sheep he sees with the intention of running to B. As he keeps well back from them,

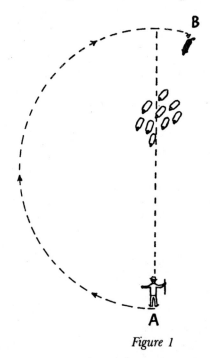

Figure 1

however, he is bound to see the farther-out sheep when he gets to about C and will cast back round them. And so on right round the whole flock which runs together towards the handler. The close-run dog probably never sees the farther-out

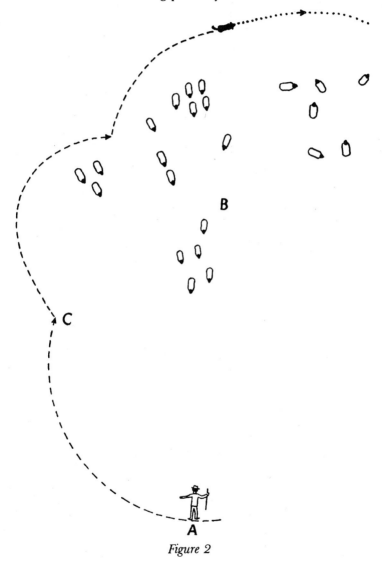

Figure 2

sheep, cuts in on the first bunch he comes across and leaves the rest. The barking dog will send the rest hell for leather to the furthest corner of the field.

Do not get the idea that all Border Collie pups will instinctively run according to my diagrams. Some will and some will not, but those which do not are usually easier to put right than common-bred pups. I have never seen a pedigree pup run barking round or through the middle of a flock of sheep as so many dogs from so-called 'good working strains' have a habit of doing.

Don't get the idea either that all you have to do is buy any pedigree pup and read the instructions in this book (or any other book for that matter) to be certain of ending up with a good dog. We live in an age of specialization. But specialists have a habit of fixing their eye on one particular goal and, like a horse with blinkers, becoming oblivious to all that goes on around them. They forget that very important word 'enough' and, in going to extremes, produce specialist animals suitable for only *one* purpose.

Trial men are no exception and, although they have not altered the appearance of the Border Collie, they have exaggerated certain characteristics to just as great an extent as the show enthusiasts (fanatics I call them!) have altered the head of the show Collie. Just as the breeder of show dogs has bred for one purpose, to win in the show ring, so the breeder of trial dogs has bred dogs for one purpose – to win trials.

Although trials have done so much to raise the general standard of working dogs they have undoubtedly resulted in the production of a great many dogs that are of no *practical* use to anyone. Many classically bred dogs have quite an abnormal instinct to work. The skilled trainer can use this keenness as a foundation for training the dog to the very highest degree. It will respond like a flash to each and every whispered command like the high-powered, well-tuned car responds to the slightest touch of the accelerator. But a high-powered car with a bad driver is far more dangerous than a low-powered!

The common-bred dog that has little inclination to work may never do much good. He is, however, much less likely to do the harm which an over-keen, well-bred one will do in incompetent hands. The very keen pup will work anything and everything that moves, but even the gamest of dogs stands a poor chance against a bus. I should say that at least 75 per cent of the *best* working dogs, had they been allowed to run around on their own (worse still with other dogs) would have worried livestock. Personally, the keener a dog is the better I like it, as I shall be explaining when we come to training. This very strong instinct, however, will only be an asset when under control. When out of control it can prove a quite unpredictable liability.

A common criticism of what practical shepherds often call 'fancy-bred' dogs is their strength of 'eye'. A great asset in moderation, carried to extremes – as I have already said – this can render a dog quite useless for practical work. Nothing connected with dogs tries my patience more than the young dog which casts out in a beautiful circle then suddenly comes across an old ewe behind a whin bush and 'freezes' on the spot. There he stays, chin on the ground, staring at the sheep and there she stays stamping her foot and staring at the dog. And there am I at the other end of the field, first coaxing the dog to come on then cursing him for not coming on, all to no effect. After a while I decide to call him back and let the sheep go, but he is in a sort of trance by now and *cannot* take his eye off her. In the end I have to go and collect him, but I know a good many farmers who would feel more inclined to go for their gun!

Of course, such a dog would be of little use in a trial. Although the uninitiated often think that well-bred dogs are those which crawl about on their bellies and clap down after every move, the rules of the I.S.D.S. state that 'excessive clapping will be penalized'. At one time regarded as stylish, trial men, no doubt having realized how impractical it was, now dislike this style of working, and at many trials a special prize is offered for the 'best upstanding style'.

In breeding for any particular object, however, one inevitably gets some that fall short and some that go beyond the ideal. A dog with too little 'eye' for trial work may be an excellent all-round practical worker. The dog with too much 'eye' is of no use to anyone except an expert, and not always of much use to him. Unfortunately these experts are masters of the art of getting a 'strong-eyed' dog to get on his feet and come on. I say 'unfortunately' as many trial winners are dogs which the average farmer with limited time and skill, not to mention patience, would never get to move. These dogs are used at stud to produce more super dogs for the experts or useless dogs for the novice. It all depends on which you happen to be!

Which brings us to the great weakness in trials and indeed competitions of any sort. Without an intimate knowledge of the history of the individual animal it is impossible to tell how much of the finished article is a result of inherent working ability and how much the skill of the trainer. In no class of training are cleverer trainers to be found than the top sheep-dog men, but they form an infinitesimal minority of the people who work sheepdogs. The result is that these experts often win the highest awards with dogs which the great majority could not work at all.

I cannot see how this weakness will ever be overcome. From experience of them all, I consider that trials held under I.S.D.S. rules are far more practical than gun-dog field trials or police-dog working trials. Any dog that qualifies to run in the International would, if trained for the job, make a good dog for practical work. The majority are, in fact, practical workers which help their masters with the everyday tasks that crop up throughout the year.

At the same time, there are undoubtedly a great many disappointments amongst novices who pay good prices for well-bred pedigree puppies. This is due, not so much to the weakness to which I have just referred, as to failure to appreciate this weakness. People pay too much attention to what they see and too little attention to what has taken place before the dog ever sees a trial course. For example, gripping is taboo in a trial. Until recently it was the only 'crime' for which the judge could stop the dog running and disqualify it on the spot. But I doubt very much if any dog that really would not grip has ever been a consistent winner in trials. Many readers will be surprised to learn that I know top trial men who, if they have a young dog that will not grip, go to a great deal of trouble in teaching him to do so. The person, therefore, who assumes that, because trial dogs don't grip, their pups won't grip is very often in for a big disillusionment.

Equally far off the mark is the person who wants a dog to work cattle and assumes that a trial dog could not do so. A stubborn old ewe stamping her foot at a dog is a quite formidable foe. Any dog that will walk straight up to her and, by sheer will-power and determination, make her back into the pen which she had no intention of entering will almost certainly face a bullock. The very fact that he does not use his teeth in retaliation to the sheep's threats calls for far more courage than if he did. In my experience the well-bred dog is just as likely as the common-bred one to turn anything that can be turned. What is more important, he will keep back from cattle instead of chasing them as fast as they will go, as so often happens with the so-called 'good cattle dog'. The advantages of this are obvious in the case of milking or calving cows. Of even greater advantage is the ability of such a dog to steady and get under control a bunch of wild bullocks or heifers which the close-run, noisy dog will make all the wilder. Photograph 5 shows Judy walking quietly behind a herd of milking cows and, photograph 6, going in to turn two that have broken away.

Not that I object to a dog that barks *provided he does it at the right time*. To force several hundred sheep into pens for dipping is a very different thing from forcing five sheep into a pen on a trial ground. It is no use the dog just staring at them, as only one or two sheep on the outside will see him. What can he do? Bark and/or bite. In such circumstances, the barking dog will do as much good as, and a great deal less harm than, the gripping one. Although many trial-bred dogs will bark at cattle, very few will bark at sheep in any circumstances. The old adage about buying a dog and barking oneself often applies literally to the owners of such dogs!

To those who want a dog to earn a little money in its spare time, so to speak, a well-bred Border Collie bitch will show a far better return on the extra capital outlay. Not only does she need to be well bred but you must make sure that she *is* registered. Many breeders of these dogs are very careless about registering their pups and, even if the pedigree is quite genuine, a pup cannot be registered unless its parents are registered. That will not make your bitch inferior as a worker but it will reduce the value of her pups.

To sum up, I am convinced that the advantages of obtaining a pedigree Border Collie pup easily outweigh any disadvantages. But the most important point, often overlooked, is to buy a pup bred from good all-round workers, not just a pup with a pedigree.

There are several other types of Collie quite distinct from the Border Collie in that they are 'loose-eyed' workers. Most of these are native to Scotland and include the old-fashioned Scotch Collie from which the modern show Collie is descended. Now practically extinct, I have clear recollections of several of these dogs in my youth and believe that, in my early efforts to walk, I was assisted by one. They were all easy-going, level-headed dogs, useful but not flashy workers, and quite willing to lie about the place when there was nothing better to do.. Personally I think it a great pity that this type has been practically exterminated by the increasing popularity of 'strong-eyed' dogs. For all-round farm work they were often far more use than the classically bred dog.

The Welsh Collie is much smaller, about the size of the average Border Collie, but very different in type. They are usually more 'common' looking, often with a 'cocky' tail, detested by all Scots shepherds I have met. My experience of them (which is not very great) is that they are hardy but often rough workers. In any case, this breed also has been practically ousted by the Border Collie.

Other British sheepdogs are markedly dissimilar in appearance from the Collie types. The best known of these are the Bearded Collie of Scotland, the Old English Bobtail and the Old Welsh Grey. The similarity not only in type but in style of working suggests that they have common ancestry. All three are close-run, 'loose-eyed', noisy workers usually with a short 'youf, youf' bark, quite different from that of other Collies. All three breeds were very popular with drovers, but as the latter have been replaced by motor transport so their numbers have decreased.

On the bracken-covered hills of some parts of Scotland the Beardy is still used as a hunter. Instead of 'wearing' its sheep to the shepherd, both get behind the sheep and the dog drives them out of the corries and from amongst the bracken, working back and forth across the hill, keeping up an incessant barking. To the trial enthusiast a crude method of working, but if anyone were to try to use a trial winner on this type of ground, not only would he lose a lot of sheep, the chances are he would lose the dog too!

The Bobtail is, or was, the dog which helped the Down shepherds with their flocks, besides being popular with drovers for both sheep and cattle. Like the folded flocks, the shepherds and the drovers, this breed has now gone out of fashion. It is similar in type and colour to the Beardy, except that it is bigger, heavier and not nearly so active. Like the Shire horse the hair on the legs is supposed to give protection when constantly working in mud. I never cared for Shires, least of all the hair on their legs! I have, however, met old shepherds who would swear that, for constant work in muddy fields, the Bobtail would see a Collie dead. Not having had the opportunity to make comparisons for myself, I leave it at that. It should be remembered that the Bobtails these men used bore little resemblance to those seen at shows, which have so much hair they look more like sheep than dogs.

Quite distinct from the above types of working dogs are the Welsh Corgis and Lancashire Heelers, both of which are essentially cattle dogs. Best known of these (as pets of the Royal Family rather than as cattle dogs), are the Corgis, of which there are two types – Cardigan and Pembroke. Having bred the latter since 1933 and having owned more of them than all other breeds put together (at one time I had twenty brood bitches), I can speak from experience. From that experience I can say that, as drovers' dogs, to keep a bunch of cattle moving on the road, push them into waggons, keep sheep off troughs while food is being put in, or push them through a muddy gate, a good Corgi is hard to beat. They are, however, close-run, rough workers and as all-rounders on the farm cannot be compared with the Border Collie. If, however, you have, or fancy, a Corgi, it is quite possible that you may find one that will be a useful asset around the farm. The best ratter I ever knew, also the hardiest cattle dog, was my first Corgi, from which all of my strain were descended.

On the other hand, you may want a German Shepherd Dog as a guard and, although I have seen many Border Collies that would put the average German Shepherd Dog to shame as a guard, there is no doubt that more people will be afraid of the latter. Some German Shepherd Dogs make excellent sheep and cattle dogs and, if you want one as a guard, it should be possible to find one to do both jobs. The chief trouble with Corgis, German Shepherd Dogs and other breeds not usually used for work is that they are not bred for work either. Although many, bred from generations that have never seen a sheep or a bullock, still have a strong herding instinct, many do not and there is no way of telling except by actual trial.

When considering the various herding breeds from which you can choose, it should be remembered that there are no clear dividing lines between those native to this country. Not that I am suggesting that British sheepdogs are mongrels. Nothing annoys me more than when my friends in the dog-show world suggest they are. As herding dogs they have been bred pure for thousands of years (no one actually knows how long) and the average show dog registered at the Kennel Club carries a far higher percentage of foreign blood. At the same time, there is no doubt at all that the various types have been crossed to produce dogs for different kinds of work and to suit individual tastes. The only attempt to produce a breed for a specialized job has been since the stud-book of the International Sheepdog Society was first compiled. Considering the length of time that sheepdogs have been in use, however, the Border Collie, as a breed, is still in its infancy.

It should not be forgotten that, in all classes of farm stock, crosses and 'hybrids' have often proved better commercial propositions than either parent. We all know, of course, that, while certain crosses appear to inherit the best characteristics of each parent, there are others which do quite the reverse. In sheep, cattle, pigs and poultry, trial and error with large numbers of animals has fairly well established which crosses are likely to be a success and which are not.

Unfortunately in dogs this experimentation has been carried out on much too limited a scale to be of much benefit. By crossing a Collie and a Terrier one might combine the former's herding instinct with the latter's toughness. One could just as easily breed a dog with the Collie's instinct to herd combined with the Terrier's instinct to kill. Just the sort to avoid. By crossing two herding breeds, however, you are unlikely to produce anything worse than either parent, which fundamentally have been produced for the same purpose.

Mongrels as a rule are not worth the food they eat, but I have known some first crosses, including some that I purposely bred myself, which were really outstanding dogs. These appear to have had the hybrid vigour usually associated with first crosses. This vigour is mental as much as physical, often producing dogs which have more intelligence and character than either parent. Of course, one does not expect to produce crosses that will win trials any more than one would expect to produce a world milk record by crossing an Aberdeen Angus with a British Friesian.

It is not a trial winner we are after, however, and, for our job, I have known crosses which were better than classically bred dogs. Although the pure Beardy has never been very popular with Scottish shepherds, the half Beardy has always been in great demand. One of these, Nell, was training when I first wrote this book and she was shown in several of the original illustrations. She was by a Beardy dog (a rough and ready worker whose ancestry I do not know) out of an exceptionally well-bred, 'strong-eyed', bare-skinned Border Collie bitch.

Nell had enough 'eye' for all practical purposes *and* she stayed on her feet and got on with the job without any encouragement. She used her brains too and, once she had done same job several times, would do it by herself. When working, she kept her mind on her job, but was quite easy to get away from it and did not go creeping off to work the hens the minute I turned my back. Although she never barked in the work for which I used her, I could easily get her to bark at cattle or sheep 'in bulk'. There are pedigree Border Collies which will do all that but they are extremely rare. My experience is that they are becoming rarer and rarer. Incidentally, I bought Nell at ten months when she was just starting to run and paid more for her than is often asked for registered pups at the same age.

I see no reason why a Border Collie cross with the Old English Bobtail or the Welsh Grey should not be as useful as that with the Beardy. I have had no experience of either, but a south country shepherd I know very well has often praised to me the virtues of the Bobtail cross. In fact, he says that he wishes he could find one now but, like many of his colleagues, always works registered Border Collies. One reason, probably the main reason, for this is the greater value of the recorded pups they sell from time to time.

It should be remembered that just as much care should be taken in breeding or selecting a cross-bred pup as a pure-bred one. The 'failures' in certain crosses are often due, not so much to the cross, as to the individual animals which have been crossed. Two animals, neither of which is considered good enough to breed pure stock, are mated in the hope that they will produce a 'good cross'. That they don't is really less surprising than if they did.

On the subject of breeding, I should perhaps mention that, in the sheep districts of Scotland and the north of England, there are many strains of good working dogs which are neither registered nor eligible for registration. These dogs are just as pure bred as any registered dog but their owners never bother about pedigrees. A pup from such a strain will cost less and probably end up just as good a worker as a pedigree one. You must be *sure* of its origin, though.

Two foreign breeds of working dogs have recently been introduced into the UK. First of all came the Huntaway from New Zealand, which shepherds in this country seem to think is either absolutely wonderful or not worth the food it eats! Of course 'hunters' were always used to put sheep out of the bracken and corries of the Western Highlands. Often Beardies or half Beardies, these dogs would simply rush back and forth behind the sheep barking incessantly until they were out in the open when a 'wearing' dog usually took over.

By its name one would expect the Huntaway to do the same thing, but it will wear too and will go out any distance. I must admit I have only seen one working but believe he was typical of the breed. He gathered a big flock from a steep Wiltshire down, barking all the time. The sheep probably lost more weight than if a good Border Collie had been working them, but the Huntaway would certainly appeal to the impatient farmer who feels he can never get the job done quickly enough. When the sheep were on the road he was in his element, and made the Border Collies working with him look a bit useless. The Huntaway is a much bigger dog than any of the British herding breeds and certainly offers an alternative for those who do not like a quiet, strong-eyed dog.

When my wife and I were in Australia in 1979 we took such a fancy to several Australian Cattle Dogs which we saw that we decided to import one. There have been several other imports and there are now quite a number of the breed in this country. Originally known as the Queensland Heeler this breed was produced by crossing blue merle collies from Scotland with the Australian Dingo, and later on a Kelpie cross was introduced. The object was to produce a dog that could stand up to harsh conditions on the huge cattle ranges; and there is certainly no other breed that can beat the Australian Cattle Dog at its job. Unfortunately very few have gone to work cattle here, but those which have, have been very successful. It is found in two colours, red speckled and blue speckled and is the toughest breed I have ever met. At the same time it is naturally biddable and easily trained. It is also extremely intelligent and with a remarkable desire to stay with its master. Unlike the keen Border Collie, which seizes every opportunity to sneak off and work sheep or poultry (sometimes belonging to other people), the Australian Cattle Dog is always there. But he is no less keen to work when there is work to do. It is essentially a cattle dog, although I have one which works sheep just as well as cattle. He is very close run and noisy but gets on well with my very tame Polled Dorset ewes. To start with he did tend to grip but, surprisingly, I had less trouble stopping this than I have often had with Border Collies. I have known others which work sheep but if they grip they bite very hard and I cannot recommend the breed as a sheepdog for a novice trainer. It surprises me that it was as recently as 1983 when the first Kelpies were introduced to the UK. This is really just a strain of Border Collie. It might be more accurate to say that the Border Collie and the Kelpie have the same common ancestors. The first sheepdog trial recorded in Australia was held in 1872 and was won by a black and tan bitch called King's Kelpie. She was out of Gleeson's Kelpie and from stock imported from Scotland, some of which were red in colour. So famous did Kelpie become and such was the demand for 'Kelpie's pups' that the strain became known as the Kelpie, which in Gaelic is a water sprite.

The Kelpie's style of working is similar to the Border Collie and in Australia and South Africa they compete against each other in sheepdog trials. Australian sheepmen, however, maintain that for a hard day's work with thousands of sheep the Kelpie is far superior to the Border Collie. The two breeds are frequently crossed and the resultant offspring are popular with sheepmen. We hear so many complaints about the modern Border Collie in Britain lacking 'power' that I would have thought a real outcross to a Kelpie might benefit the breed enormously. However, that is only my opinion.

At this stage I might repeat that I have been discussing the choice of a young puppy. A young dog that is shaping to work can be judged partly on its own merits which makes its breeding that much less important. A dog's merits can also be assessed to some extent 'on inspection'. This is nothing like as good a guide as the dog's breeding, and that can be misleading at times. The easiest way to judge a dairy cow is to measure the milk in the bucket and the easiest way to judge a working dog is to see it working. But you cannot do either if the cow is not in milk or if the dog has not started to work. In both cases one can, if one knows what to look for, at least sort the good from the bad, even if one cannot say how good or how bad it will be.

In judging a working dog, I always start with the head which is, after all, the place where the brains are kept. It is significant to note that all wild dogs, gun dogs, sheepdogs and the vast majority of mongrels throughout the world have fundamentally the same shape of head. A Spaniel and a prick-eared Border Collie may look very different but the skeleton of the skull with ears and lips removed would be very similar. The only heads which are very different from that of the wild dog are to be found on dogs which have been bred purely for show. The most important point arising from this observation is that the shape of head gives no clue at all as to whether or no the dog has any instinct to work.

The head does, however, form quite a good guide to both intelligence and temperament. All the most intelligent dogs I have known (I have never met anyone who has trained a lot of dogs who disagreed on this) have had heads of similar shape. They had what I call normal heads, free from exaggeration. The skull was fairly broad, slightly wider at the back than at the front with a well-defined 'stop' (where the skull joins the foreface). Sometimes, but by no means invariably, there was a pronounced occiput or 'bump of wisdom'.

Of greater importance than the actual shape of head is the dog's expression. This is governed almost entirely by the eyes. It has been said that the eye is the mirror of the soul. There are eyes, I am afraid, which give a somewhat distorted picture but in spite of that the dog's eye is the best and often the only guide to its character. In any dog the first thing I look for is a bold, *honest* eye that looks straight at me. Usually, depending on the purpose, I prefer a friendly eye but, if not, I like it to be honest – to let me know that it dislikes me. The sort I detest is the furtive, shifty eye that won't look at me. That, of course, is a characteristic handed down from the wild dog, and here we come to something which can be, and often is, misleading – colour of the eye.

By comparison with the domestic dogs most wild dogs have light eyes, sometimes almost colourless. This gives rise to the belief that a light-eyed dog (in breeds like the working Collie that have not been bred for show quite a number do have light eyes) is treacherous. I have owned and known many such dogs and have found that this type of eye is very often misleading. Dogs possessing it are no more liable to be treacherous than many other dogs and I have usually found them intelligent and extremely alert. What I dislike about a light eye is not that it is a distorting mirror but that it is no mirror at all. I have found it impossible to judge character through a really light eye and equally impossible to judge a dog's thoughts.

Experienced trainers get into the habit of watching (often unconsciously) changes in the dog's expression from which they can form a fair idea of what is on his mind. They can tell whether he likes or dislikes something and whether he refuses to do something because he does not understand or because he does not want to do it. By watching his expression we can often see that he is thinking of doing something before he does it, enabling us either to correct or reward him as he does it. But a light-eyed dog always looks the same, which is the chief reason that I dislike them.

That, of course, is the extreme to which I said we should never go and, at the other end, is the black eye so popular amongst the dog-show fraternity. This eye is just as expressionless as the light one and it does not indicate a bright, wide-awake sort of temperament. On the contrary, I have found that it is often associated with a dull, stubborn mentality. Leaving aside those extremes, a small, dark eye usually indicates a harder, sometimes tougher, dog than the nice, kind, brown eye. The little piggy black eye often denotes cunning rather than intelligence but is rarely found in any of the herding breeds.

Ear carriage can alter considerably a dog's expression, or at least *appear* to alter it, which can also be most misleading. A prick-eared dog always looks more alert than a drop-eared, especially on first meeting it. I have not found that the shape or carriage of the ears is in any way connected with the dog's mental characteristics. One might expect a prick-eared dog to have more acute hearing and some people claim that it does. I have found that Spaniels, Salukis and even Poodles with so much 'wool' in their ears hear just as well as Alsatians, Corgis or prick-eared Collies.

In judging any animal, I like to be able to tell what sex it is from the head alone. This is perhaps more important in a working dog than in any other animal. For work I never object to a 'doggy' bitch, and several of the best I have ever known had very masculine features. I have yet to meet a 'bitchy' dog, however, that was worth the food he ate.

Getting back to the type we are looking for, there is no point in having a head, no matter how brainy it is, without a body to

carry it. The working Collie has a body built for speed and
stamina far above the average for dogs of the same size, and
there is a very interesting point connected with this. By breeding
dogs to work, breeders have produced a body physically capable of that work. It is indeed very rare to find a working
sheepdog which lacks the necessary speed, stamina and agility
for the job. Breeders of show dogs, on the other hand, have
concentrated on producing a body which they *believe* is ideal
for the work in mind. I reckon that about 50 per cent of dogs
produced in this way are nothing more or less than cripples,
and the number which could (assuming they would) stand up
to a hard day on a rocky mountain in Scotland or Wales is
infinitesimal. The moral in that is to choose a dog that moves
well rather than one which theoretically is built on the right
lines to move well.

Generally speaking, a working dog should be built on the
lines of a good hunter – head and neck nicely laid into (not
just stuck on) the shoulders which are well laid back on well
sprung, deep ribs. Plenty of length from withers to loin: loin
short, strong and firm. If you press down on the loin of a well
built, fit dog it will spring back up as the hand is released.
Powerful hindquarters (a horse or a dog pushes itself forward
from behind, it doesn't pull itself from the front) with good
second thighs and hocks down near the ground. Plenty of
good, clean, but not excessive, bone. Strong feet with well-
arched toes and well-developed, firm pads. The most important point is that it should move easily and lightly on its feet. A
Shire horse takes a longer stride than a Thoroughbred but it
won't win the Grand National!

I dislike a dog or a horse that turns his toes out in front and
have usually found this fault associated with weak pasterns.
Toes that turn in like a Foxhound nearly always go with strong
pasterns and, in spite of the show-ring aversion to them, I
have never found that they do any harm. That is, of course,
provided that it is the toes which turn in and not the elbows
which are out.

One of the biggest differences between a horse and a dog is
that, when trotting naturally, the dog carries its head much
lower – usually in line with the back. In the case of the Collie,

the tail is also set on lower and should be carried low. This tail carriage is a somewhat controversial subject. As I have already mentioned, Scottish shepherds detest 'cocky' tails. It would appear that their Welsh colleagues do not care for them either, as the Border Collie, with its low-tail carriage, has practically ousted the Welsh sheepdog with its gay tail.

At the same time there is no doubt that a gay tail denotes a gay temperament and many dogs which always keep their tails down do so because they lack the 'guts' to put them up. But too gay a temperament may give us a dog that is a careless, rough and often an erratic worker. Many dogs that carry their tails up when not working, and especially when playing with other dogs, drop them the minute they see a sheep. Dogs, of course, tend to carry their tails much higher than bitches and little puppies always carry theirs up. Although I consider a nice low tail always looks better, I have not found that a gay tail does any harm unless the dog works with it up.

Some people think that a big, strong dog is necessary to work cattle. That this is a fallacy is proved by what I have already said about Corgis as cattle dogs. My father always kept that type and some of them could literally hang on to a bullock until they pulled him to the ground. What they would have done to a sheep I shudder to think! I have known many small dogs which, by using their brains rather than brute strength, would have persuaded the bullock to turn without having to grip at all. A big dog is often clumsy and he may lose more than he gains by the extra size.

Whether to have a dog or a bitch worries some people – usually quite unnecessarily. The question is to a great extent answered by what I have already said about castration and spaying. But it must not be regarded as the complete answer for all lazy and irresponsible dog owners.

The trial-bred Collie is the only breed I know where the working instinct has been developed to such an extent that it is often stronger than the sex instinct. I have seen a dog working alongside a bitch in season without so much as looking at her. A well-bred young dog is therefore just as liable to go looking for sheep as for bitches. The latter of course do not usually present any great problem even when in season. Dogs from surrounding farms are the problem, creating as they roam much bad feeling between neighbours. So much so that, if you definitely do not want to breed from your bitch, it may be worth having her spayed.

On the whole, bitches tend to be more amenable, and therefore easier to train, while dogs are usually bolder and less 'touchy'. But anyone who expects every bitch to be amenable and every dog to be bold is in for a disillusionment. In working Collies more than any other breed if I find a pup I like at a price I can afford its sex does not worry me unduly.

When buying a dog, it is just as important as when buying any other animal to make as sure as you can that it is healthy. Dogs are probably prone to more ailments than any other class of livestock. This is not, as some people think, due to inbreeding for show points. There is certainly more disease amongst show dogs, but I believe that is due to their being exposed so often to the risk of disease rather than to a tendency to contract diseases. My experience, over many years running a fairly big kennel of all sorts of breeds, is that all breeds are susceptible to disease when they meet it. The most susceptible of all is the shepherd's dog off an isolated place where it has had no opportunity to build up a natural resistance.

The right thing to do is to have a dog examined by a Veterinary Surgeon before you buy it. Few people do this, but the Vet's examination fee will be much less than his fee if he has to attend an ailing dog when you get it home. Apart from any illness a dog may have, a good Veterinary Surgeon will spot defects in conformation, eyesight, etc., if they happen to be there. As with any other animal, good health is reflected in a bright eye, glossy coat and a general liveliness, especially in a young animal. But a coat will not look healthy unless it is clean and, as remarkably few working dogs are clean (even those appearing before the general public in trials and demonstrations) it is unlikely to be much of a guide.

The curse of all dog breeders and many dog owners is virus diseases. At one time the commonest was distemper, but this was superseded by the more virulent hard pad. Now we must add parvovirus and leptospirosis, and it is likely that there will be others in the future. Dogs do not get colds like horses or human beings and any tendency to 'gummy' eyes or nose is almost invariably due to one of the above infections. No one with kennels would dream of bringing such a dog on to the place and I advise anyone not to consider such a dog without first having it examined by a Vet.

Unfortunately, virus diseases often affect the brain, sometimes leaving permanent nervousness or mental deficiency. In severe cases these are easy to detect, but sometimes disease affects the brain very slightly, when it is extremely difficult to notice. That is, until you start trying to train the dog, when you will soon realize that there is something lacking. Unless you are very lucky your Vet is unlikely to be of much help to you here. I am not suggesting that all dogs which have had hard pad or distemper will be affected mentally. Having had more than my fair share of experience of these diseases I know that some are and some are not. All I want to emphasize is the importance of making sure that the one you buy is not, as this acquired weakness will handicap a working dog just as much as if it were hereditary.

Perhaps next in importance to be on your guard against in buying a dog are skin troubles. Unhealthy looking bare patches may be due to (1) follicular mange, not contagious but difficult, often impossible, to cure; (2) sarcoptic mange, very contagious but usually easy to cure; (3) eczema, often caused by wrong diet; (4) ringworm, very contagious to both animals and humans; (5) worms; (6) fleas and/or lice or (7) generally poor condition.

You will see, therefore, it is sometimes possible to transform a dog from a miserable-looking creature to a picture of health, by giving it a good bath and feeding it properly. In another, which does not look any more unhealthy, this may be quite impossible. The only person who can diagnose skin trouble with any degree of accuracy is a Veterinary Surgeon who will examine a skin scraping under a microscope. Most farm-reared puppies have worms, fleas and/or lice. This is by no means as inevitable as some people seem to think, and could invariably be prevented by a little care and attention. External parasites are easy to see and to eradicate and I would never turn down an otherwise healthy puppy because of fleas or lice.

Roundworms can seriously retard the growth of young puppies but it is astonishing how quickly they will pick up after dosing. It is also astonishing the masses of worms that can live in such a small body. In adult dogs it is tapeworms that are usually found. These seem to have little effect on some dogs but a very serious effect on others. They can usually be eradicated quite easily, but a dog running on a farm is almost certain to become reinfested, making periodical dosing necessary. Apart from the dog's health, it is the larvae of this worm which causes gid or sturdy in sheep.

Rickets is another ailment fairly common amongst puppies. A young puppy that shows a tendency to rickets by being slightly back on its pasterns will usually recover completely by being put on an 'anti-ricket' diet. I doubt very much, however, if a puppy which is really rickety at, say, four months, will ever have the speed, stamina and constitution to make him a good, sound working dog.

The prospective puppy buyer should be aware of possible hereditary weaknesses. Progressive Retinal Atrophy (PRA) has been known in Border Collies for a very long time and used to be quite a problem. In affected animals the eyesight is usually normal until about five years of age when they go blind, sometimes quite suddenly. Since the ISDS adopted a policy of only registering puppies from parents which had been tested 'clear' for PRA the problem has been considerably reduced. Tests cannot be carried out until a dog is two years old so that you cannot have a puppy tested and must hope that because the parents are 'clear' the pup is OK – which it probably will be.

Hip Dysplasia (HD) is not common in breeds kept solely for work for the simple reason that an affected dog cannot work and is therefore unlikely to be bred from. This deformity shows on X-ray and many breeders have their breeding stock regularly 'hip scored'.

Deafness, either total or unilateral, is more common than is generally realized and in a recent survey it was found that 36 breeds were affected in the UK. It is of interest to readers because it has been found in Australian Cattle Dogs and Border Collies. In the latter it is usually those of the blue merle colour which are affected.

It hardly needs saying that a totally deaf dog is useless as a worker and a unilaterally deaf one is severely handicapped. Although it can hear in one ear it has no idea of where the sound is coming from. Modern technology has made it possible to test puppies for deafness by the BAER method at 4–6 weeks old. If there is a suspicion of deafness in a strain it is well worth having a puppy tested before buying. It is not easy to identify a deaf puppy by observation and almost impossible if it is unilaterally deaf.

Part II

PRACTICE

5

Principles of Training

WE NOW come to what some of my readers will consider to be
the most important part of the book but I do not think it is.
The first step in putting anything right is usually to find out
why it went wrong. The good mechanic can see something
going wrong and put it right before any damage has been
done, while the bad one will wait until the machine is wrecked.
But the reason why the clever mechanic can do that is because
he knows *how* the machine works and therefore knows what is
likely to go wrong.

I cannot emphasize too strongly that dogs are not machines,
but the reason I have written so much in Part I is to help you
to *prevent* things going wrong. To be able to do this, the first
essential is to acquire some 'dog sense'. This is the ability to
understand the canine mind and to be able to see things from
the dog's point of view. Without some dog sense the advice I
have still to give you will be of very little value.

The art of training is nothing new; it is as old as the domestic dog itself. But, as in everything else, one generation learns
from the mistakes and discoveries of its predecessor, so that
better and easier methods of doing the same thing are
devised. This applies to the training of dogs no less than to
anything else. By using up-to-date methods results can often
be achieved in hours which would previously have taken
weeks. Apart from which, up-to-date methods usually give far
greater pleasure to both teacher and pupil than do the old
fashioned.

One must not forget, however, that although the ways of
applying them have improved so much, the basic principles of
training any animal have never changed. It should also be

remembered that the mentality of the dog has changed very little since prehistoric man first took it into his care.

Having already dealt with the mentality, let us now be clear on these basic principles. First of all it should be repeated that a dog is not 'almost human' and I know of no greater insult to the canine race than to describe it as such! The dog can do many things which man cannot do, never could do and never will do. I mention this again because many people, in trying to see the 'almost human' qualities which are *not* there, completely fail to appreciate and utilize many of the canine qualities which are.

At the same time, bear in mind that we can do a great many things which dogs cannot do. No matter how good a dog is, without a master to guide him, he is of as little use as a hill shepherd without a dog. There is no use, therefore, buying a trained dog and expecting him to do everything you want, unless you are able to let him know *what* you want.

The first and probably the greatest difference between the human and the canine mind is that dogs do not reason as we do. Amongst those who have studied the subject there are great differences of opinion as to whether dogs reason at all and, if so, to what extent. There are some who go so far as to say that a dog does not work anything out in its mind; it does not think about what it has done or what it is going to do, only about what it is doing.

With this I entirely disagree, and believe that dogs, or at least *some* dogs, think a good deal more than we imagine and in much the same way as we do. As dogs cannot talk (thank heaven!) we can never be certain on this point. Therefore, the point cannot be proved one way or the other. Some dogs, I believe, will work out the best way to tackle a problem. Unfortunately, dogs seem to be much more adept at working out problems for their *own* benefit than for the benefit of their masters! If the master allows him, however, a keen working dog will often find some way of doing a particular job which his master never thought of.

I had an outstanding example of this with Floss, my first sheepdog. An old black-faced ewe had strayed with her lamb and, seeing her some distance away, I sent my bitch to fetch her. The lamb, however, was quite young and the ewe faced the bitch. Knowing that, if she could not bring the ewe to me, she would stay with it till I got there to help her, I went on 'looking' the rest of the sheep. Suddenly I heard Floss barking, which surprised me as she had never barked at a sheep in her life. On looking towards them I saw that the ewe, protecting her lamb, was chasing the bitch round in circles and she, in her turn, was trying unsuccessfully to chase the sheep towards me. To my great surprise she suddenly gave up her efforts. Instead, she alternately barked at the ewe and ran away. This intrigued me and I left her alone to see what she would do. Every time the sheep stopped, Floss would turn round, bark in its face and rush off *towards me* with the ewe in pursuit apparently butting her in the rump at times.

In this way she brought the sheep a distance of about 500 or 600 yards in quite a short time. When she reached me she kept up this teasing operation round and round me and, by that time, the old ewe was so mad that it did not realize that I was there. She came so close that although I was standing in the middle of a forty-acre field I had only to put my hand down to catch her. From then on, if a single sheep faced Floss, she would bark in its face and get it to chase her to me.

I have never seen or heard of another dog adopting these tactics, which were entirely against the dog's natural instinct to 'wear' (very strong in this particular case). I do not, for a moment, think that Floss said to herself 'If I bark in this old so-and-so's face it will chase me towards him.' She started barking in sheer exasperation because the sheep would not move. Many silent working dogs will bark at a sheep that faces them. When the sheep started chasing her, however, I feel she must have realized that she could get it to me by running in my direction. From that point her intelligence and her powers of reasoning combined to overpower her tremendously strong 'wearing' instinct. She also remembered how to overcome this problem and in future would get the sheep to chase her straight to me without first chasing her round in circles.

The many accomplishments of Floss and other dogs I have owned or known have convinced me that *some* dogs do *sometimes* reason to a certain extent. Nevertheless, there are certain points which must always be borne in mind. (1) Only a minority of dogs are capable of working out problems in their own mind. (2) Of those which do, most of them do so to suit their own ends, developing in the process cunning rather than intelligence. (3) Only when there is complete understanding between master and dog (a much-too-rare state of affairs) will the dog work out problems for the benefit of his master. (4) Dogs can be, and often are, trained to such a high standard (e.g. for trials) that they become entirely dependent on the commands of their master and will not *attempt* to use their own brain, far less succeed in doing so.

All training must, therefore, be based on the assumption that *dogs do not reason*, which we might call principle No. 1 – a principle, incidentally, which is accepted by practically all successful trainers of all classes of animal.

The second point to remember is that, no matter how much some owners would like them to, no dog 'understands every word that is said to him'. It does not understand any words at all. It merely understands sounds and it is just as easy to teach a dog to lie down by saying 'get up' as by saying 'lie down'. Neither does it understand whistles, as some people imagine.

If dogs cannot reason or understand words how can we make them understand what we want them to do? Dogs, in common with all animals, learn by association of ideas. That simply means that they associate certain actions with pleasure or displeasure. Quite naturally they tend to repeat the action which provided the pleasure and to refrain from doing that which caused the displeasure.

We too have associations of ideas, and I believe these are fixed in the dog's mind in much the same way as they are in ours. For instance, when I hear a pipe band my memory is immediately taken back to the agricultural shows I attended as a small boy. In those days both my father and grandfather were successful Clydesdale exhibitors and these shows were *the* events of my life. At practically all of these the pipe band was the only arena attraction, apart from the parade of livestock. Although I have heard the same music on many and diverse occasions since, it always brings back pleasant memories of those agricultural shows.

There are others, however, to whom this same music brings back memories of very different kinds, sometimes very unpleasant. And there are some – mainly from south of the border – to whom it is just a horrible noise to be avoided as much as possible! What I am trying to explain is that the same thing, in this case a sound, may be associated by different people with entirely different experiences and is sometimes not associated with anything at all. Exactly the same applies to the dog.

The strongest and most lasting associations are created by terrifying experiences, very pleasant experiences and experiences that are happening over and over again, in that order. First experiences always create stronger and more lasting associations than do subsequent experiences. If, for instance, a pup gets kicked by a bullock the first time it sees one, it may never go near one again, but if it gets kicked after it has learned to work, then it will merely be more careful in future. For the purpose of training, we create the pleasant and unpleasant experiences by what is known as correction and reward. We cannot tell the dog what we want him to do as we could a person, but we can make it very pleasant for him to do what we want and equally unpleasant for him to do what we don't want. Eventually he should learn right from wrong and you will be able to tell him what you want him to do.

A good comparison is a child who has never seen an open fire. Provided he is old enough to understand, he can be told that, if he puts his finger in the fire, he will get burnt. Provided that he respects and trusts the person who tells him, it is very unlikely that he will try it out. A younger child, however, cannot be made to understand that the fire will burn, and must be kept away from it. If, however, a baby accidentally does put his finger in the fire it is unlikely that he will ever do so again. Probably he will avoid the fire, at least for a very long time, because he associates it with severe pain. In time the child will be old enough to be told that the fire will not hurt if he does not get too close. The dog never gets to that stage. We cannot tell a dog that, if he does what we want, he will be rewarded, and if he does what we don't want he will be corrected. We can only let him do it and then reward or correct him.

As we do this we try to get the dog to associate a certain sound (whistle or word of command) or movement (hand signal) with the action we want. For instance we say 'down', push the dog down into the position we want (correction) and then praise him for lying there (reward). Soon, often incredibly soon, the dog on hearing the word 'down' will lie down without any correction at all. To the dog, the word 'down' is merely a sound which we teach it to associate with lying down. As I said, it would be just as easy to teach him to lie down by saying 'get up' so long as we always used the same command. Many shepherds teach their dog to lie down to a low 's-s-s' but, to the same sound, I teach my dogs to attack.

Once, on entering a ring to give a demonstration of obedience, I overheard one member of the audience saying to another, 'Listen to what he says. We may learn some of the secret words.' This remark showed a complete lack of 'dog sense'. The same applies to those people who think it is terribly clever to be able to control a dog by whistle only. There are others who think that a sheepdog can *only* be controlled by whistle. On a hill some of the control must be by whistle as it carries farther than the voice, but it is rarely that a farm dog is as far away as that. With the exception of some Welshmen who can whistle as well as they can sing, practically all trial men use words of command as much as whistles.

So far as you are concerned the first thing to decide is what commands you are going to use – and stick to them. To control a dog entirely by whistle you must be able to whistle the same note in different tones, as if you were talking to the dog. And you must be able to do it with many different notes for different commands. Few people can do that. I certainly cannot, and I therefore rely chiefly on words of command which come naturally in varying tones according to what I want the dog to do.

In my advice on training I shall frequently use these commands but, in doing so, I am not suggesting that they are the best or only commands that can be used. If you are training a dog for the first time and have never used any commands before, you might as well use these – unless you don't like them! But, if you have got into the habit of using a certain command for a certain action, and, more important, if your dog responds to that command, then stick to it.

One command which is practically universal is a long shrill whistle to drop a dog at a distance. This I advise you to use for several reasons. Being shrill it carries farther, more so to a dog than to us, as a dog can hear a note so high that it is inaudible to the human ear. Very often, when you want to stop a dog, he is rushing about in excitement and cannot or will not hear you shouting, even if he is quite near. You therefore want a high-pitched, penetrating sound which will *make* him hear.

Many Border Collies 'clap' to a whistle when they have never been taught to do so. This is probably because (a) they have been bred for so long to 'clap' to a whistle that it has become more or less an acquired instinct or (b) they don't like the high-pitched sound and cower from it instinctively as they do from a harsh tone of voice. Or, of course, it could be a combination of the two. Whatever the reason, I have found that Border Collies will 'clap' to a whistle more readily and more quickly than to any other command – which is as good a reason as any for using it.

There is on the market a 'silent' whistle used quite a lot by gun-dog trainers. This is so high-pitched that we cannot hear it, although the dog can. I have tried this whistle and found that some dogs hate it, giving the impression that it hurts their ears. Anyhow, I like to hear myself whistling, and, not being able to whistle with my fingers, use the old-fashioned hill shepherd's whistle. This is simply a piece of metal bent over with two holes in it. There are also more modern plastic whistles which stay in the mouth without being held by hand.

They are placed against the underside of the tongue and, by blowing gently, one should, after some practice, be able to produce a long shrill blast. With much more practice you may be able to play a tune on it!

What do I mean by correction and reward? By the former I certainly do not mean belting a dog with a stick which even in these enlightened times far too many farmers are still inclined to do. Correction is, perhaps, a misnomer as it can refer to any action on the part of the trainer to make the dog do something. It can mean simply pushing the dog into a sitting or lying position or it can mean really correcting it for something like grabbing hold of a sheep.

There should be a very clear dividing line between correction and punishment. Correction is something which *has* to be done to get the dog to understand what he must do, and should, therefore, always be used to the absolute minimum. Punishment, although it may take the same form, is administered only when the dog has been deliberately disobedient, and can, therefore, be quite severe. That is if you are *quite* sure he has been disobedient. No dog can be disobedient until he knows what he *should* do, but many are punished simply because their owners have failed to make that clear. Many dogs are punished for trying their very hardest to please a master they cannot understand. Remember, too, that what might be very severe punishment to one dog might have no effect at all on another.

In the training of sheepdogs I rely almost entirely on two forms of correction. The first I use when it is possible to get hold of the dog, and I vary it according to how severely I want to correct him (which depends more on the temperament of the dog than on the 'crime'). Sometimes I simply go up to the dog, grab him by the scruff with one hand and give him a mild shake. At other times I get hold of the dog with both hands and, holding him by the loose skin under the neck, lift him straight off his feet (photograph 7). Staring into his face I then really shake him, sometimes until he must be nearly giddy. Between these two extremes I use varying degrees of correction all on the same principle.

It is important to remember that one of the reasons why man is able to train animals is that, because of his ability to reason and their inability to do so, he can give the impression that he is much more powerful than they. When a dog is firmly held by the scruff it *cannot* retaliate in any way, and can, therefore, be shaken mentally far more than physically. The object of all correction and, in fact, all training, is to apply it to the dog's mind, not his body. For that reason this form of correction is the most effective I know. Don't forget, however, that it is also quite the most severe (excluding of course the brutal methods which I should so much like to apply to those who use them).

8 and 9 Until it has learned its 'sides', a young dog should always be sent out from behind the handler

10 Make the best of any opportunities to provide work for a young dog

11 So long as the dog just lies there quietly, sheep like these Dorsets will go on grazing

12 But if she gets to her feet and walks on quietly, the sheep should move towards the handler who can walk backwards encouraging the dog to 'wear' them to him

13 'Way t'me' — dog and handler both move anticlockwise

14 'Come bye' — dog and handler both move clockwise

2 Judy 'eyes' something she cannot see

15 and 16 Teaching a dog to walk straight forward and stand,
using a line

To grab a sensitive young dog and give him a thorough shaking may completely shatter his confidence and, as I shall be emphasizing as we proceed with training, get to know your dog *before* correction is necessary.

To correct a dog when he is some distance away the obvious thing to do is throw something at him. The 'weapon' which is usually already to hand is a stick and there must be very few shepherds' dogs that have not had one thrown at them in the course of their career. I cannot recommend it. I *have* thrown a stick at a dog, but it is an extremely dangerous practice even when great care is exercised – and there is never time to be careful. A far better weapon is a handful of loose earth or gravel but these are usually difficult to find in the middle of a grass field.

I once read an article on training gun dogs and since then have used an extremely simple 'corrective instrument'. It is merely a piece of half-inch or three-quarter-inch rubber hosepipe about two feet long. This can be thrown much farther and with far greater accuracy than a stick, it cannot injure the dog and it can be used to hit the dog if he is near enough to do so.

However, I rarely hit a dog, and only in certain circumstances. If two or more dogs start to scrap I would hit each of them in quick succession (hit them hard), very probably breaking them up before they got 'stuck in' to a real fight. If a young dog gets hold of a hen and starts to 'pluck' it I would, if I could, get hold of him and shake him as already described. If I couldn't, I might still get near enough to 'belt him one' while he still has hold of the bird. This is much more effective if the dog doesn't see you throw the hosepipe – it appears to be something which has descended upon him from Heaven. If a dog has had one or two 'doings' for catching hens you won't get near enough either to shake him or hit him, but you can throw this 'weapon' and, with luck, may hit him while he still has hold of the bird.

Reward is much more self-explanatory but there are many ways of providing it. Because of the submissive instinct most dogs like to be fussed and petted, and this can be used as a means of rewarding them for doing something right. You may have heard that petting spoils a dog but, like so many other theories, it is expounded only by those who have never tried it. Petting and slobbering over a dog whenever he wants petting can do no good at all and may do endless harm. To praise a dog by fussing and petting when he has done something you want him to do goes a long way to encourage him to do the same thing again. It is too little correction, not too much praise, which usually spoils dogs. To reward as much as possible achieves the same objective more quickly and much more pleasantly for both dog and owner than to train by correction only.

The sheepdog and gun-dog trainer is much more fortunate than those who try to train companion dogs to be obedient. If he starts with a well-bred dog, keen to work, it will find its own reward in its work. It is for that reason that sheepdogs *can be* and often *are* trained without any word of praise from the trainer. If trainers of dogs for circus, or many other, purposes were to work on the same lines as many sheepdog and gun-dog trainers, their dogs would be pictures of misery. It is only because they get so much pleasure from their work that many sheepdogs and gun dogs do not look just as miserable.

Too little attention is paid to the importance of correcting and rewarding by tone of voice and expression only. As you proceed with training the dog learns to associate the way you look at him and speak to him with doing the right or wrong thing. The untrained dog, however, will instinctively react to these 'aids' – an asset which should not be thrown away.

This is a good example of how easily man can assume the role of canine pack leader. Always having a pack of dogs of widely varying types and temperaments I can best illustrate this with an example.

At one time we had two dogs which wanted to be leaders, one Fahmi, an old Saluki, the other, Jason, an exceptionally powerful young Maremma Sheepdog (bred more as guards than herders and, on their native hills, capable of killing a wolf). Fortunately all my other dogs were quite happy as members of the pack. We had to exercise the greatest care to see that Jason and Fahmi did not fight as the former would almost certainly kill the Saluki which had been with us much longer and had no intention of 'standing down' in favour of the young

dog. To both these dogs, my wife and I were their only superiors, they would not fight if we were there, and we worked them
together with the rest of the team.

Now we come to the interesting and important point in that
neither of these dogs ever attempted to fight with any 'member
of the pack', nor did any of the latter ever offer to fight with
them. We had several which would scrap amongst themselves,
but they would never square up to Jason or Fahmi. This was
not due to any mysterious 'dog language' or telepathy; it was
due merely to the superior dog's attitude of approach, his look
and his tone of voice. Once, when I was exercising Jason with a
tiny Papillon in London's Regent's Park, a Poodle chased the
little bitch. Jason immediately galloped towards his 'property'
to intercept and face the Poodle. He stopped dead, pulled himself to his full height and fixed the poor Poodle with an icy
stare. I could almost feel the unfortunate intruder wishing that
the ground could open up to swallow him, and, tail down, he
quickly slunk off to his master. Jason nonchalantly turned and
strolled towards his little friend, who by now was rushing
towards him and would, if she could, have flung her arms
round his neck in gratitude.

That is not an 'almost human' story. It is, in fact a very
canine one, the sort of thing that many animals would do
under similar circumstances. If the Poodle had not made himself scarce, Jason would have growled at him and, had he
shown the slightest inclination to retaliate, he would have
attacked. In other words he would, with a minimum of fuss,
have shown the other dog who was boss. If would-be trainers
would only adopt exactly the same policy they would find that,
in nine cases out of ten, physical correction is quite unnecessary.

Of course a 'look' or a rating by harsh tone of voice has little
or no effect on a dog like Jason. He was the defiant, pack-leader
type and, great dog that he was, I often had to use every ounce
of physical and mental power at my disposal to make him realize that *I* was boss. Such dogs, however, are in the minority,
especially amongst the true herding breeds. Most dogs, like
most people, are not only willing but anxious to have someone
to tell them what to do. As well as being able to correct a dog by

staring at him and using a harsh tone of voice (the equivalent of a dog's growl and not to be confused with shouting) one can also praise a dog by smiling at him and speaking in a friendly tone. What one says matters not the slightest, it is the way it is said that counts.

If there is a secret in training I believe it arises from the ability to apply the type of correction and reward suitable to the particular dog, to strike a balance between the two and, most important of all, *to apply them at the right time*. The majority of failures in training are due to (a) too much emphasis being put on *how* to correct or reward a dog, and too little on *when* to do so, and (b) trying to work on the dog's body rather than on his mind. The minds of some dogs can only be got at by making disobedience a painful occupation. But unless, in his own mind, the dog understands quite clearly *why* he is being corrected you will merely be causing unnecessary suffering. Apart from which you will almost certainly make the dog distrust and dislike you. The best servants are those who work for a boss because they like and respect him, rather than because they are afraid of him.

Because we are trying to work on the dog's mind it is essential that correction or reward is applied when that mind is bent on doing right or wrong, irrespective of what the body is doing. As a common example of this, take a dog that chases a rabbit. You curse him in a very angry tone, but in the untrained dog, this form of correction is not nearly severe enough to control the strong hunting instinct which makes him want to chase. He may be so intent on chasing that he does not even hear you. Eventually the rabbit goes to ground and the dog starts digging at the mouth of the burrow. Not being a terrier, however, he will soon give that up and come back to look for you.

Now, there are two courses you could have adopted. (1) You run after the dog and, grabbing him while his head *and* his mind are half-way down the burrow, give him a thorough scolding and shaking. (2) You stand where you are and wait till the dog comes back and then give him a good hiding. If you adopt the first course you will have corrected the dog for chasing rabbits. If this is the first rabbit he has chased you may have gone a long way to curing him for life. If you adopt the second course, you will have corrected the dog for coming back to you. Not having corrected him for chasing rabbits he is just as likely to chase the next one he sees. But having been severely corrected for going to his owner he is going to be very wary of doing so in future.

It is astonishing how many dogs suffer unnecessarily and are completely ruined by the above tactics. How many dogs which, guided by their own instincts, go off after a bitch in season or go rabbiting with another dog are punished for coming home? Two dogs I knew, belonging to the same person, used to go off on regular hunting expeditions in the woods and, in the opinion of the owner, were always punished for doing so. Did this punishment do any good? Of course not. Why should it?

I have seen these two dogs sitting about for a whole day in a wood just behind the house. They had been hunting and were obviously tired and more than anxious to go home. But they had been punished so often for going home (not for hunting) that they were afraid to do so. Eventually hunger drove them home to take what was coming to them for staying away so long. Having had one or two, I know just how aggravating is a dog that disappears into the wood the minute your back is turned. A dog that goes off and comes back, however, is infinitely preferable to one that goes off and has to be brought back.

Returning to the question of correcting or rewarding when the dog's mind is on doing something, it should be remembered that, *before* my hand picks up the pen with which I am writing, my brain has told it to do so. Before the young dog actually moved in the direction of the rabbit its mind was on doing so. It may have given quite some consideration to chasing the rabbit, which would have been obvious to the observant trainer. Had you, during this period of indecision, scolded the dog severely he might not have chased the rabbit at all. If, as he set off in pursuit, you had thrown something at him and had been lucky or clever enough to hit him, he would almost certainly have stopped in his tracks.

If, however, you wait until he is down a burrow, then approach with the intention of grabbing him as already mentioned, he may decide to come back. Although he may not have started to do so, to correct him after he has made that decision to return is to correct him for coming back. It is often far better to wait for a second opportunity to correct a dog while he is really in the act of doing something wrong than to risk correcting him when it is too late. It should be noted that I am referring to the untrained dog which has not yet learned right from wrong.

If you do catch the dog with his mind down the burrow you should, as you shake him, use a word of reprimand such as 'No' in a very angry tone. The dog should then associate this word in such a tone (itself a form of reproof) with correction. Next time a rabbit pops up, providing you are quicker than you were last time, you can say 'No' just before the dog sets off and the chances are he will stop. For this, of course, he should be well praised and, again, a word of praise, such as 'Good dog', used. The word you employ does not matter (I know some excellent sheepdog trainers who use words not usually published in books!) but it is important always to use the same one. I always say 'No' or 'A-a-ah', a sound which can be made in a very angry tone. Eventually the word of correction or reward can be used in varying tones, the importance of which should become apparent as we proceed with training.

So far, we have been dealing with the associations which we try to build up in an effort to get the dog to do what we want and, just as important, not to do what we don't want. We must, however, always be on our guard against undesirable associations which we may unintentionally create or which may be formed by circumstances outside our control. One of the best illustrations of this was an experience of my own. We had two young Salukis which galloped for the sheer joy of galloping. Round and round the place they would go like Greyhounds round a race track, usually pursued at varying distances by a variety of other breeds. One day, in the course of this game, they raced right round our biggest enclosure of about seven acres and back through a four-foot gate into a very small enclosure by the kennels, followed by our big Maremma, like a Shire horse trying to keep up with a pair of Thoroughbreds!

As luck would have it, the Saluki bitch, still going flat out, was coming out of this small enclosure just as the Maremma reached the gate. My wife and I both saw what was going to happen but there was no time to do anything and I reckon that the combined speed when they crashed must have been in the region of 40 m.p.h. The Maremma's shoulder went into the bitch's flank like the wheel of a car and, quite frankly, we thought she was done for. But Salukis are far stronger dogs than most people imagine and, although both Tessa and Jason were very stiff for a few days, neither suffered any ill-effects. Physical ill-effects that is. Jason is one of those dogs which crashes his way through life and, like many of the thumpings I have given him, he had forgotten the crash in five minutes. Not so Tessa, an animal with a very different temperament, and in spite of our thankfulness that she was so well physically, we were much concerned as to the effect it would have mentally.

Sad to relate, the beautiful Tessa was killed on the road about a year after the above incident – the only dog we have ever had which was killed by a car. But she never went through that gate if the Maremma or the Saluki dog who was galloping with her were anywhere near. Neither would she gallop with them in the big enclosure, and if either approached her from behind she was liable to panic and bolt for home. Obviously she associated not only the Maremma with this terrifying experience but also the Saluki who was galloping with her, although he had nothing whatsoever to do with it.

That did not surprise us greatly as we had seen similar examples. What did surprise us was that when, only about a week after this episode, we took our team out on a demonstration we found that this bitch was only too willing to gallop and play with both the Maremma and the Saluki. This showed that she associated the experience not only with the two dogs present but also with the place where it happened. Just how much she did associate the place we discovered when we found that she had no fear of either dog in an adjoining enclosure entered by an adjoining gate. The two gates shut on to the same post, and I must say that it really surprised us to find that this bitch would quite happily go through the one on the left with both dogs, yet refuse to go through the other if either dog were present.

The point I want to emphasize in that example is that this

bitch associated this experience with both the place *and* either of the two dogs mentioned. She would go through the gate quite happily on her own or with any of our other dogs. She would go with either of the dogs concerned anywhere except in the vicinity of this gate. She did, however, associate the gate and the dogs together with a most unpleasant experience and was terrified of its happening again.

Fortunately that did not face us with a very big problem and Tessa might have forgotten in time. Quite a lot could have happened which would have been much more serious to us. My wife and I were both fairly near when it happened and our first worry, after we saw that she was not injured, was that she might associate either or both of us with the crash. There was, in fact, almost as much reason why she should do so as that she should associate the Saluki dog who was playing with her. Since then, a friend of mine was standing near his puppy when it touched an electric fence. For months afterwards he could not persuade the puppy to come near him although it showed no fear of fences. Unfortunately, most unfortunately, it associated its owner, not the fence, with this frightening experience.

Returning to Tessa, the most likely consequence might have been that she would be afraid of the Maremma dog at all times instead of just in the vicinity of the accident. Or she might have associated it with a gate and been afraid to go through any opening at all like a gate. And so we could go on. So many different associations can be built up round the same experience that it is impossible to foretell what the effect will be.

These, of course, are terrifying experiences, the sort of things one always hopes will never happen. We should, nevertheless, always remember how easily and quickly they *can* happen. There is much less likelihood of that sort of thing happening in associations which we deliberately try to create in training. It can, however, happen there too, especially in the use of what I call 'shock tactics'. If, for instance, your dog prepares to chase a rabbit and you throw something at him and hit him, the chances are at least ten to one he will rush straight back to you. The reason he does so is because something which descended upon him from Heaven frightened him and he rushed to you for protection.

If, just before you threw the missile, you called him in a firm tone he will associate that word in that tone with this unpleasant experience. On hearing the same command in future he will probably come to you before you throw anything. *But* if he looks round just as you are throwing something he will see you do it and, instead of rushing to you, may quite naturally run away. Worse still, if he turned to look at you it is almost certain that he had decided to come to you and, as I have already explained, you would scare him of coming to you, not of chasing rabbits.

Because of its importance I have dealt at some length with association of ideas. Many people say to me, 'How do you train all these dogs? I expect it's just patience.' Well, it isn't *just* patience, essential as patience is. The *first* essential in training is will-power combined with an active mind. This enables one to concentrate, to anticipate what the dog is going to do and, therefore, to correct or reward him *as he does it*. Most people who cannot train animals are unable to do that. They are like those people who find themselves sitting on the ground wondering what made their horse shy, when a good horseman, realizing that it was going to shy, would have prevented the horse doing so. The ability to concentrate and to act quickly cannot, unfortunately, be acquired through reading a book, but it can be improved with practice – *and the will to do it!*

To be able to work a dog successfully, it is a great help to understand bovine and ovine, as well as canine, mentality. Many people think that sheep are stupid, thereby divulging their own stupidity. Anyone who has had to match his wits against an old blackfaced ewe who has made up her mind to get into a field of roots knows just how clever she can be. Not once but many times have I seen an old ewe watch someone mend a hole in a fence and immediately go to see if she could break it open again. Not only did she go back to the exact spot where she had previously got out – she had stood there watching with the obvious intention of going back.

Intelligent animals, like intelligent children, are almost invariably those which get into mischief. The only animal I know to equal a pet lamb on that score is a hand-reared kid. The sheep and the goat are about equal in intelligence. The 'silly goat' phrase has been coined by ignorant people. My wife and I have trained several sheep and goats to do tricks, including all the sheep for the very successful series of TV commercials advertising wool. We have found that their standard of intelligence is as high as that of the horse. I have even found them much easier to train than some dogs. They cannot, of course, be taught to do many of the things that dogs can be taught, but that is due to their having different instincts, not lack of intelligence.

Why then will hundreds of sheep run from one dog of about half the size of any one of them? They run from the dog just as instinctively as the dog runs after them. The common idea that the one is due to stupidity and the other to intelligence proves how many people are unable to differentiate between the two. The whole of Nature is made up of hunters and hunted. Domestication has not killed the sheep's instinct to escape from a beast of prey any more than it has killed the dog's instinct to hunt that prey.

It is of interest to note in this connection that sheep will run from a cat. This is not, if you reflect, nearly so unreasonable as some people when their instincts get the better of them. My own wife, who will tackle the most ferocious Alsatian, is petrified of a mouse! This in spite of the fact that she is able to reason and knows perfectly well that the mouse is just as anxious to escape from her as she is from it.

It is to our great advantage that the sheep, the least domesticated of all domestic animals, has retained this strong instinct to escape from a dog. Otherwise the production of mutton in sheep-farming districts would be an uneconomic proposition. Of course, the strength of this instinct, like all others, varies between individuals and one does find some sheep (more often in certain breeds) which will not run at all. Speaking generally, however, the only instinct in the sheep which will overpower the instinct to escape from a dog is the maternal instinct.

Although I have seen it so often, I have never ceased to marvel at the difference between a ewe before and after lambing.

The timorous creature that will back away from a dog is, in a matter of minutes, transformed into an aggressive and quite formidable foe ready to give its life in defence of its newborn lamb. Fortunately this instinct does not last long. As the lamb becomes stronger and more able to escape the ewe relies more and more on her ability to escape with it and less and less on her ability to protect it.

Cattle, of course, are very different. Being a much bigger and stronger animal, the cow is much more able to protect herself, and therefore, much less dependent on her ability to escape. But a very small dog can still control a herd of cattle. In this respect the relationship between the dog and his charges is very similar to that between master and dog. He is endowed with far greater intelligence than the cow (cattle are far less intelligent than sheep), with a set of teeth capable of inflicting very severe pain and with far greater agility of both mind and body.

If a dog has a weak submissive instinct and refuses to try to learn, we must teach him by correction and reward that it is to his *own* advantage to do so. If cattle refuse to run from the dog instinctively, he is capable, by adopting the same principle, of teaching them the same lesson. The same, of course, applies to sheep which have become too tame to run instinctively from a dog. It is for that reason that many dogs which are excellent workers of wild hill sheep are no use at all with cattle or park sheep. They are like the people who can successfully train an easy dog but lack the strength and agility of mind to tackle a difficult one.

There is another similarity in the relationship between the dog and his charges and the master and his dog. Many disobedient dogs are afraid of their masters and many dogs which are not afraid are disobedient because they lack respect. But all the best dogs are those which are fond of, *and* respect, their masters. Some dogs will chase cattle or sheep, inducing a state of terror, while others are so 'soft' that the tables are turned and even sheep will chase them. But cattle or sheep which are worked regularly by a good dog will lose their fear of that dog (not necessarily of all dogs) without losing their respect for him. This is an all too rare state of affairs attained only by the right sort of dog which has had the right sort of training.

Although animals can be trained by correction only, the best results are invariably achieved by correction *and* reward. The dog can very easily correct the old ewe which faces him by nipping her on the nose but he cannot give her a friendly pat and tell her she is a good girl when she does right! He can, however, leave her alone, which to a sheep is a very good form of reward. The dog, like the trainer who commands respect without fear, is the one who never applies correction unless it is necessary. When he does, he applies it quickly, at the right psychological moment. More important still, when it is done it is *done*.

The trouble with many dogs is that when they are challenged and retaliate they lose their tempers and do not know when to stop. Need I add that that is far too common a fault amongst 'trainers'? The dog which chivvies and hurries his charges, like the owner who nags at his dog, builds up fear without respect. Stock which are regularly worked by a dog which keeps well back and leaves them alone, but which will whip in like lightning to correct any that stray, will soon learn to walk quietly in the direction the dog 'tells' them to walk.

6

General Care

Feeding – Housing – Illness

FEEDING

LET US now imagine you have a new pup of about eight weeks old. Perhaps I should repeat that it is not essential to get a pup of that age, but I shall start there for the benefit of those who have done so. In the strict sense of the word, you are not going to start training this pup for some time but you can, in bringing up a pup, do a great deal to make subsequent training easier – or more difficult. At this age the pup's chief concern is to eat, sleep and play, so we might start with feeding, housing and exercising.

It is over thirty years since this book was first published, during which time many changes have taken place in agriculture. Unfortunately one still sees many farm dogs covered with lice and fleas and riddled with tapeworms. They are either chucked any leftover scraps or, if lucky, given a bowl of flaked maize and skimmed milk. It has never ceased to amaze me that these dogs actually do a day's work. But a correctly fed dog is far more efficient than a badly fed one. He is also less likely to become ill; or if he does, is far more likely to make a good recovery. And he wi'. last longer, an important factor when good dogs are hard to find and expensive to buy. To get the best returns from his beef cattle a farmer will feed for optimum weight gain and his dairy cattle will be fed for maximum milk output. So, surely, it makes sense to do the same with the working dog?

It pleases me to note that the number of farmers who treat their dogs badly appears to be decreasing. Perhaps they are better educated and perhaps the fact that good dogs cost a great deal more than in the past contributes to the change. It is certainly very welcome and I hope is continues.

In 1960 I recommended raw meat as the staple diet for all dogs, and at that time our own dogs were fed on virtually nothing else. But, as I say, times have changed! Gradually raw meat has become more and more difficult to obtain – and more and more expensive. As labour costs increased, the time spent preparing the meat would add to the cost, which would have been even greater if we had cooked it. So eventually and very reluctantly we changed over to proprietary dog foods; and I must admit our dogs look just as well and are just as healthy as before.

It is worth noting that in some countries it is illegal to feed uncooked meat to dogs. By 1994 these restrictions have spread and there is every indication that they will continue to do so.

Whether one cares for modern methods or not, dogs are certainly much easier to feed than in the past. And scientists have, at last, done some research into the nutritional needs of the dog. This brings it into line with other farm stock for which balanced rations were the 'in thing' when I was at college – a long time ago! A great deal of research on the nutritional requirements of the dog have been carried out by the National Research Council (N.R.C.) of the National Academy of Sciences of the USA and by the Animal Studies Centre at Walton on the Wolds, Melton Mowbray, Leicestershire. The result is that we can now work out balanced rations for dogs in much the same way as has been done with other farm stock.

Dogs are, by nature, carnivores, although most modern dogs would, more accurately, be termed omnivores. But whether working dog or pet dog they all need certain nutrients, such as carbohydrates, proteins, fats, vitamins, minerals and trace elements. The diet should contain these nutrients in the correct proportion. The National Research Council has worked out the requirements for many types of dog under many different conditions. Most reputable commercial feeds are based on the results of these findings. Although good old-fashioned wholemeal biscuit meal and biscuits can still be bought, there are an ever increasing number of complete and mixer diets on the market. There are canned meat products, semi-moist foods (which do not need to be refrigerated as they have added preservatives) and dry foods. The latter can be flaked, expanded or extruded. Most can be bought more cheaply in bulk and, kept in cool, dry conditions, dry diets can be kept for some considerable time. As I said, any complete diet from a well-known firm will have been worked out to the N.R.C.'s recommendations and no extra vitamins etc should be added, as this could do more harm than good. But some dogs dislike a completely dry food and the addition of gravy or milk will not seriously upset the balance of the diet. Some firms market special diets for working dogs. The N.R.C. recommend the normal protein requirement at around 20 per cent and that of fat 5 per cent. But dogs in very hard work, such as sheepdogs on steep hill farms, will not only use up more energy, but will also need up to three times the amount of food needed by a similar sized guard dog doing routine patrol work. Because it needs so much more food it will also need that food to be in an easily digestible, less bulky form. This can be achieved by adding more fat and protein.

Apart from the amount of work the dog is doing it should be remembered that, unlike other farm animals, dogs are not bred with a view to economic food conversion. They are bred to work and vary considerably in the amount of food required for a maintenance ration. An energetic, excitable dog may well require twice the amount of food of a rather phlegmatic one doing exactly the same work.

If you prefer to mix your own diet, a mixture of three parts rice, bread, or cooked grain or biscuit meal to one part animal product such as fish, meat or eggs provides a basic diet, but this may need the addition of calcium and vitamin D. The recommended quantities are 1 g. bone meal and 15 IUs vitamin D to 100 gs. dry food. The dog in hard work can usefully have more animal product and less cereal. Cheese, meat, eggs, fish, kitchen scraps such as cottage cheese, yoghurt, peas and beans (both high in protein) will help. Cooked carrots or potatoes can also be fed but do not contribute much to the diet except bulk. Raw eggs were at one time said to be dangerous but I always fed them with no adverse effect. Most experts now agree that a raw egg a day will do no harm. Cooked eggs are certainly more easily digested by the dog, but if raw eggs are easier – feed them. Incidentally, in addition to the excellent food value of the egg itself the shell can also provide valuable calcium, if your dog will eat it. If you are making up a dry mix it can be moistened with milk (whole, skimmed or calf rearing, whatever is available).

A working dog is best fed twice a day, say a third of the total food before work and the rest when he has finished for the day. But never feed a tired dog immediately he has finished a hard day's work. Allow him time to relax, have a small drink and then feed him an hour or so later. This way he will feel more like eating and will be able to digest his food much better.

Our own dogs have always been fed six days a week with a dog biscuit on the seventh. Carnivores do not by nature eat the same amount of the same food at the same time every day. They gorge themselves on what they have caught (if it is big enough), then sleep it off and go off to catch something else – if they are lucky. We have never found that dogs benefit in any way from regularity of feeding. Indeed, we believe that if they are not fed regularly they do not fret if they cannot be fed at exactly the right time – as so often happens with a working dog. And if you change the brand of food every now and again your dog is less likely to go off his food if you cannot get the usual one. Apart from which dogs like a change as much as we do.

If you have a bitch and intend to breed from her and she is on a good diet up to the N.R.C. requirements, she should not need any extras. Feed her normally for the first five or six weeks and then gradually increase the diet by about 10 per cent per week, until she is having half as much again when she is due to whelp. Also split the food into several small meals, especially if she is carrying a large litter. After whelping, depending on the size of the litter she may need up to three times her normal diet, gradually decreasing again until the pups are weaned at about six weeks. Any properly fed bitch should come off her pups weighing approximately the same as when she was mated. The frequently heard remark that 'she looks poor because she has just reared a litter' is simply an excuse for bad feeding.

Our pups stay on four meals until they are about eight weeks old and then, depending on how they are doing, are reduced to three. They usually stay on three meals until about six months and then have a light milky feed in the morning with the main meal at night, until they finish growing and get over the lean 'yearling' stage. We have found that a puppy which is not very keen on its food often eats up better if the number of meals is cut down to let it 'feel the bottom of its stomach'.

If you are buying a puppy of about eight weeks the most sensible thing to do, and best for the puppy, is to ask the breeder for the diet sheet – in fact any good breeder should just give you one. Stick to this for a week or so. The pup will be under enough stress leaving his litter mates and moving to a new home without adding even more by upsetting his stomach with different food. If you want to change his diet do so gradually over a few weeks. A puppy of eight weeks should be having three or four meals a day, decreasing to two by the time he is five or six months old.

To finish this advice on feeding, if you are a mere male and have a wife, mother, sister or any other female who wants to feed the pup, let her do it. It is almost certain she will make a far better job of it than you will. Do not get the idea that the pup will become attached to her in preference to you. If a dog will leave the person who works it and go to the one who feeds it, it is either a poor dog or has a poor trainer. The late David Dickson of Hawick, a very good friend of mine and breeder of several National and International Champions, never fed his pups. He never even took them out until they were ready to work. But they did not leave him and go home to his niece who fed them. Not the ones he kept, anyhow!

We have been talking about food but we must not forget water. Food is of secondary importance to water – which is cheaper too! A dog can exist for weeks without food, but only for days, or under some circumstances, hours, without water. Looking at it from a practical point of view the efficiency of a working dog can fall by 50 per cent if he is denied access to water for any length of time. Water is needed by the body for many functions including temperature control and as an aid to digestion.

Most working dogs can quickly find their way to the nearest water trough or stream, but if you are working in unfamiliar country, or country where you know water is in short supply, take some along with you. And never forget that if you are feeding a dry diet without adding milk or water, water *must* be available *at all times*.

HOUSING

The dog is an easy animal to house. With the cat it shares the distinction of being the only animal which man has allowed, since prehistoric times, to share his own home. Not all men, of course, and it is a strange fact that amongst those who deny the dog this privilege are to be found those who are most dependent on him.

It is said by some, including some very successful sheepdog trainers, that allowing a dog to live indoors makes it soft. The kids spoil it, the wife spoils it and when you want it to work it runs off home. In the first place, amongst those who believe this, I have not met one (and I have met many) who had kept working dogs indoors to prove his contention. In the second place, even if the wife and the kids do 'spoil' the pup, if he leaves his work he must either be a poor dog or have a poor trainer.

The whole of the dog's mentality is against such a thing happening. As a puppy he will want the sort of affection which he is more likely to get from a woman than a man. But all sons of fond mothers do not turn out to be 'cissies'! As a puppy he will also want to play, preferably with 'children' of his own species, but human ones make a good substitute. The dog's submissive instinct, however, makes him want a leader, a master, who will give him something to do. Alongside this instinct develops, or should develop, the herding instinct. By the time he is ready to work, these two instincts will combine to nullify any attachments he has for the family.

Even if the submissive instinct is weak, the keen young dog will invariably attach himself to the person who lets him work, due to association of ideas. He associates reward in the form of work with the person who works him and, therefore, sticks to that person in the hope of further reward. In the process, the submissive instinct should develop, and the dog should accept this person as a master – if he behaves as a master should!

Theory is of little value without practical evidence to back it up. Going back to the days of my youth, my father always kept a Collie outside so that it would not become soft. But he always let me have a Terrier which lived in the house. Now, as I have already mentioned, these Collies were tough but they had nothing on some of the Terriers I can remember. Nor were they any more obedient – often a good deal less so.

In training difficult dogs I sometimes come up against a kind of invisible barrier. I progress so far, but no farther. The dog does what he is told – unwillingly. Something is lacking – call it co-operation, understanding or what you like – which means that, although he works *for* me, the dog never works *with* me. This sort of barrier divides two individuals, human and canine. It may not remain if the same dog is taken over by a different person or if the same person takes over a different dog. But, if the dog and the person are going to get any pleasure out of the partnership, it must be broken down.

The best, and indeed the only satisfactory, way I have found to break down such a barrier is by taking the dog indoors to live with me. Living with a dog is like living with a person – you either grow to understand and like each other or you realize that the sooner you part company the better!

In a moment of weakness I once accepted as a gift a very big three-and-a-half-year-old Alsatian bitch which was under sentence of death as being quite untrainable. Having fought and struggled with her for about two years I had her more or less under control. I gave police-dog displays with her but she never worked *with* me. A moment's relaxation on my part, and she was liable to vanish into the crowd!

She had a mania for rabbiting and, if I allowed her any freedom at home, she was *gone*! Hours later she would return, often so exhausted that she could scarcely stand. My wife, who has a very good 'command' of dogs, could do nothing with her. If I were away from home she exercised Quiz on a plough line but soon found that, if a rabbit was within scenting distance, this great black brute would jerk her off her feet and run away either from or with her.

Well, as I say, that went on for about two years, until we lost an old Alsatian bitch that had always lived indoors. With some apprehension we decided to try Quiz as a house dog. We need not have worried. Almost overnight she became a different animal. She did not give up all her bad habits just like that but, quite suddenly, I realized that the barrier had gone. I was able, for the first time, to get at her mind. For the first time she was willing, even anxious, to do what I wanted. She remained a 'hard' bitch, requiring plenty of work, and a good deal of will-power to keep her under control, but she was no longer difficult.

Quiz became one of the best-known demonstration police dogs in the country and made many TV appearances. More important, she became devoted to me. In contrast to the early days when all her efforts were directed to getting away from me, she eventually refused to leave me. If I were writing and my wife took all the dogs out for a walk, Quiz would rush out with the others, then realize that I was not going and come back in to lie at my feet. If a rabbit popped up in front of the pack I could call her back while all the others went after it. When I was not there my wife had no trouble controlling her and did, in fact, give a police-dog demonstration on her own.

Did Quiz become soft through living in the house? I'll say she didn't! She developed a strong guarding instinct, of which she originally had none, and only an idiot would have tried to enter the house when she was in it. She would tackle anything on four legs or two, guns, sticks and all. I once saw her grab a sow by the root of the tail and throw it on its broadside, a feat I would have considered impossible for any dog.

My own views on keeping a dog as a member of the family are supported by those of the Metropolitan Police which, as they concern a large number of dogs, are all the more valuable. London's police dogs at one time lived in kennels but, after some experimenting, they now go home with their handlers. Off duty they are allowed to play with the children and more or less lead the life of any pet dog. Far from spoiling the dogs as workers, it has been found that they improve by leading a 'civilized' life. The dog becomes more attached to the handler and the handler to the dog, thereby creating a better understanding and stronger bond between the two. The result is a much more efficient man/dog team.

Of course, circumstances may make it impractical to keep your dog indoors. A common, and often insurmountable, obstacle is an unco-operative 'lady of the house'. I doubt if it is ever worth the effort of trying to overcome this obstacle! If I were a dog and had to choose between living in a shed and a house where someone did not want me, I would certainly choose the shed.

Which does not mean that a dog – any dog – deserves to live under the disgusting conditions of so many farm dogs. If a dog does his best for you, it is surely not too much to expect you to do your best for him. If he does not do his best, then get rid of him and find another that does. Provided he has a comfortable bed, a dog will be quite happy in a stable, barn or any other building on the farm. The big snag is that these buildings are usually used for other purposes and it is certain that someone will let the dog out. It is, therefore, usually worth the expense of providing the dog with a kennel of his own.

This need be neither large nor elaborate and, with very little conversion, I have used several second-hand poultry houses as kennels. The essentials are the same as for housing any other animals. It should be weatherproof, free from draughts, free from damp, well ventilated, well lit, and easily cleaned out. Inside should be a bed, a wooden box or a barrel on its side are all right, or a bench raised well off the floor. Most proprietary-made kennels have runs attached. If your dog is usually with you, this is not necessary and, in some cases, not even desirable. Some dogs will not settle in a run where they see what is going on. Put the same dog in a house where he cannot see and he will often settle down right away. This I have found applies to working Collies more than any other breed.

A young puppy should not, of course, be expected to trail around the farm all day. Nor can he be expected to grow into a strong, healthy dog without fresh air and sunshine. A run is, therefore, essential for a young puppy, but this can be a temporary affair of Weldmesh panels, corrugated iron sheets or any other available material. A roll of Weldmesh in a circle will stand on its own and I have found it invaluable as a portable run for all sorts of young animals.

Many farm dogs are kept on a chain, a practice you may expect me to condemn. A lot of rubbish is talked and written about chaining dogs – it is cruel, it makes dogs vicious, etc, etc. I have seen many vicious dogs that were chained up. But it is not chaining them up that makes them vicious. Not letting them off the chain is what does it. Whether or not it is cruel depends, not on the dog being chained, but on *how* it is chained. An intelligent, active-minded dog, given nothing to do, will become frustrated, difficult and often vicious. That applies to dogs kept in idle luxury by the dear old ladies who condemn chaining, just as much as to those that are chained.

A great deal of suffering is caused to dogs that are never on a chain. Provided a dog has a comfortable bed, a not too heavy chain of reasonable length attached to a comfortable collar and, provided it has plenty of time *off* the chain, there should be no harm in the practice. My own dogs are kept loose in kennels with runs attached and I believe they prefer that to being tied up. I have, however, kept many dogs of many breeds on a chain and, apart from rubbing the hair off the neck, I cannot honestly say that I have noticed any bad effects. Perhaps I should add that I would never tie a puppy up on a chain.

If he is on a chain, a dog can, with safety, occupy the corner of a stable, barn or other building. Alternatively, he can have an outdoor kennel. When I see some poor brutes tied up to orthodox 'dog kennels', I wonder how anyone can claim that we are a nation of dog-lovers. With a little human thought and consideration, however, a dog can be comfortable and happy chained up to a kennel. And probably a good deal healthier than shut in some dark shed. The designers of some of these kennels would appear to have gone to considerable trouble to make them as uncomfortable as possible. The opening, which should be no larger than necessary, should be on the side, so that the dog can go inside and lie round the corner out of the wind. The roof should extend far enough to form a canopy to prevent rain blowing in the entrance. A door should occupy the whole of one end or side so that it can be easily cleaned out.

Whatever type of kennel you decide upon, do give some consideration to its siting. Do not, for instance, put the back of the kennel against a wall with the entrance exposed to the north wind. It will not occupy any more room kept out from and facing the wall so that the latter provides a windbreak. If possible see that it is in a position where it will get maximum sun, but where the dog can also lie in the shade if it should be too hot. See that it is well up off the ground and on ground that is not liable to puddle. Concrete is not expensive to lay, and a slab extending just beyond the radius of the chain is well worth the trouble. In spite of all that is said against it for animals to lie upon, my own kennel runs are all concreted. I have been unable to find any ill-effects, nor have I been able to find a practical substitute. It is certainly infinitely preferable to having dogs lying about in mud as so many have to do.

A great deal of suffering, and even loss of lives, has been caused to dogs through carelessness and lack of consideration in chaining them up. Many chains are strong enough to hold a horse and, indeed, I have often seen a dog on a plough chain. How would you, at least four times as big as a dog, like to spend most of your life with such a chain hanging to your neck? If it served any useful purpose it would not be so bad. The weakest link in most chains is the spring hook, and if the dog were to give a good jerk he would very often break it. But he has still to drag this great heavy chain around for no reason at all.

A welded link chain of twelve gauge is heavy enough for any Collie. To give reasonable freedom it should be at least four feet long and it *must* have a swivel that *swivels*. It is surprising how many do not, due either to faulty construction or rust. I always like two swivels, one near each end of the chain. If a spring hook is used it should be a good one but it is not always necessary. I rarely work a dog in a collar and, therefore, if he is

tied up, I unbuckle the collar instead of unhooking the chain.
This means that the chain may be fixed permanently to the
collar. Alternatively, you can simply use a short piece of chain
as a collar. This is put round the dog's neck and the spring
hook attached to both end links. One link can then be
released, leaving the other on the hook like this:

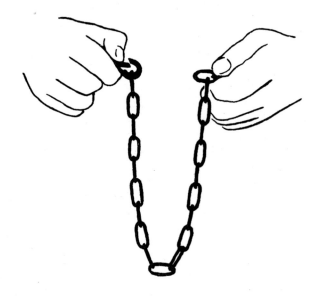

Do not fasten the chain to the kennel. Fasten it to a peg far
enough in front of the kennel to prevent the dog winding himself round it but near enough to allow him to go in and turn
round comfortably. And drive the peg right home to prevent
the chain getting wound round that. A running wire gives
more freedom on a shorter, therefore lighter, chain. Sometimes it is more convenient for a dog to occupy a long, narrow
strip than a bigger circle. There are various pulleys for fastening
a chain to a wire, but I have not found them any better than
simply putting the wire through the end link of the chain. If
you use a wire see that there are stops to prevent the dog
winding himself round his kennel, or a tree.
    Like the chain, the collar should be no heavier than
necessary. The strongest and best-wearing collars are those

with the D on the end, through which the strap is passed before being buckled. Some dogs learn to slip a leather collar. In that case I use a chain slip collar as generally employed in obedience training as a 'safety device'. I do not like leaving a dog tied up in a slip collar for long, so I put it on with the leather one and fasten the chain to both. If he slips the leather collar he is still on the chain collar. Sometimes a dog will get out of a slip collar but it is sheer luck. I have never known a dog learn how to get one off. Once a dog has slipped his collar several times only to find that he is still fastened he usually gives it up as a bad job.

ILLNESS

Nearly all dog books have a chapter on 'common ailments' or 'veterinary advice'. The best piece of veterinary advice I can offer is, if your dog is the least bit off colour, *take him to the Vet*. Over and over again I hear remarks like 'He hasn't been quite himself these last few days. I expect he has eaten something that disagrees with him.' Or 'He seems to have a touch of cold.' A few weeks later I hear that the poor dog has died, or months later it develops mental weakness, the aftermath of a mild attack of hard pad or distemper.

If you really know your dog it will be obvious to you that he is not quite his usual self long before specific symptoms can be seen. Take his temperature and you will probably find that it is up. A dog's normal temperature is around 101.5° and if it is over 102.5° take him to the Vet – *now*. Many diseases, incurable once they have a real hold, are easy to cure if prompt action is taken. Far too many people wait until the animal is dying, then call the Vet, apparently expecting him to perform a miracle. When he does not turn out to be a magician his reputation in that particular direction is often damaged considerably.

7

The New Puppy

Settling Down – House Training – Coming When Called
Chewing – Collar and Lead

SETTLING DOWN

FOR the benefit of those who intend bringing up a dog as a civilized member of the family, some advice on how to get him to behave as such might prove helpful. Circumstances alter cases, so I shall start with the puppy which is going to be brought up in the house. You can then adapt this advice (probably more helpful to the farmer's wife than to the farmer) to suit your own circumstances.

As with most other undertakings it is a good idea to make some preparations beforehand. Some sort of enclosure is a great help, big enough to allow a reasonable amount of freedom and strong enough to keep in the puppy. This pen need not be large or elaborate; the puppy will not spend its life there. A few Weldmesh panels can easily be joined together to form a collapsible pen which can be moved indoors or out. In the pen should be a bed – a box on its side is ideal. The advantages of this pen I shall be explaining as we proceed.

The first, and probably the most important, fact to remember is that, although far more intelligent and capable of fending for itself than we were at eight weeks, your puppy is still a baby. All young creatures, on two legs or four, have certain things in common. Firstly, of course, they want plenty of nourishing food, plenty of sleep and to be kept warm when asleep, although, if healthy, they will never catch cold when running about. They also want freedom to run and develop their muscles, but they should never be forced to run until exhausted. They want freedom to develop their brain, too, and at this stage it will develop much better if you *allow* rather than try to *make* it do so.

Above all, they want a protector to whom they can run in time of danger, real or imaginary. Man has taken the dog from Nature and you have taken this puppy from its mother, brothers and sisters. You have, therefore, taken upon yourself the role of mother, guardian, pack leader or however you care to regard it. The puppy will look to you for affection, protection and guidance, and do not forget that, in the natural state, dogs do not do what they like, they obey their pack leader and your puppy will actually *want* you to tell him what to do.

What I wish to emphasize most of all at this stage is the importance of building up confidence and mutual trust between you and the puppy. The first essential for success in training, and real pleasure for both dog or owner, is that they trust each other. Once a dog's confidence is shaken, as frequently happens quite by accident, it can be regained only by showing the dog that you really meant no harm, which takes time and patience. Far better, therefore, not to do anything likely to lose the puppy's confidence.

A common problem arises on the first night a puppy goes to a new home – and howls! Should you scold it, leave it alone, or feel sorry for it and take it to bed with you? If you scold it you may shake its confidence and, anyhow, who would scold a baby for being unhappy? If you take it to bed it will be happy, and tomorrow night will howl in anticipation of the same reward. The best thing is to leave it alone, when it should very soon give up howling as a bad job. You should, of course, make it as comfortable as possible. Many puppies seem to miss the warmth of their brothers and sisters as much as the company. A hot-water bottle wrapped up in a blanket often has a miraculous effect in getting a puppy to settle down and go to sleep. Often it helps if one stays with the puppy, stroking it gently until it goes to sleep, then creeping quietly away.

Remember that this puppy is going to associate you and your home with either pleasure or displeasure. Although the infant mind is less retentive than the adult's, the first impression a puppy gets of its new home is going to stick in its mind much more than those which come later. For that reason it is most important not to apply any form of correction, or do anything which may frighten the puppy, until it has settled down and accepted you as a friend. It may do that right away, but I am sure that many nervous dogs are made so during the first week in a new home.

That is not intended as an excuse for bad temperaments, which would be the same no matter how the dogs were treated. There are, however, many not-so-bold dogs which would, with proper treatment, develop confidence as they grew older. Instead, what little confidence they have is knocked out of them by people who think they must start 'training' their baby puppy the moment they get it home.

So far as a young puppy is concerned: (1) prevent bad habits developing and encourage good ones when they begin to appear, which may be at six weeks or never at all; (2) with the aid of food young puppies can be taught to sit and lie down, but it must be regarded as a game; (3) a dog is never too old to learn, but you will only create trouble for yourself and suffering for your dog if you let him become set in his ways, and then suddenly try to change those ways.

Here we come to the first advantage of the playpen. There are few households where anyone has the time to keep a constant eye on a puppy, and little ones are adept at getting under one's feet, often being trodden on accidentally. Some puppies do not seem to mind and soon learn to keep out of the way, but others will associate the unpleasant experiences, especially if they occur soon after arrival, with the new owner, the new home or with human beings in general. These risks can easily be avoided by putting the puppy in its pen when you are busy, where he can relax and go to sleep (most important in a baby puppy) instead of being constantly chivvied around.

If a puppy gets into mischief and you do not correct it, you may be allowing a bad habit to develop. If you correct it before it has gained confidence in you, it may never gain that confidence. You will, by having the puppy safely out of the way, avoid two risks – the puppy getting into bad habits or losing confidence in you. Another point is that by giving a puppy a bed in its pen it will come to regard that bed as its own property. When, later on, you take away the pen or move the bed, it is quite likely that the puppy will go to bed without any training. Even if it does not, it is much easier to teach a pup to go to its *own* bed – and stay there – than to teach it to lie in a bed it has never seen before.

Sometimes it is more convenient to have an outside pen or kennel of some sort. A convenient outhouse will do, or a roll of Weldmesh with a chicken coop inside makes an ideal pen in summer. Any puppies we intend keeping are taken indoors as much as possible. But we do not leave them in the house when we are outside as, left to their own devices, puppies can do an incredible amount of damage in an amazingly short time. Nor do we have them in all night until they are quite clean in their kennel. When there is no one around to keep an eye on him, the puppy is put out in his kennel where he *cannot* do any damage and where he does not have to be continually scolded. This system has one great advantage in that the puppy grows up accustomed to being shut up and left on its own and is unlikely to present any trouble in that direction later on.

Let us now imagine that you have just arrived home with your puppy. It is more than likely that it will show signs of bewilderment in its new surroundings and perhaps be inclined to scuttle into a corner, or under a chair, at the sound of unusual noises. At this stage the important point to remember is to *let* it, rather than try to *make* it, get accustomed to its new surroundings and to members of the household. Whatever you do, *do not chase it*. Right from the start get the puppy to come to you and see that it always associates coming to you with plea-sure. Any healthy puppy enjoys tit-bits but even more does a puppy, especially a strange and bewildered one, want affec-tion.

In spite of all that some working-dog men say about spoiling dogs and making them 'soft', baby puppies love to be cuddled and I have yet to find that it does them any harm. At the same time, many puppies are encouraged to be afraid by owners who sympathize too much in all their 'little ordeals'; who unin-tentionally praise the puppy for being afraid. Without being unsympathetic, always try to give the impression that there is really nothing to fear.

Any untrained animal has certain instinctive reactions to

other animals, including ourselves. Our intentions for good or evil are transmitted to an animal chiefly by three different means – the tone of voice, the touch of the hand and the movement of the body or hand, especially in approaching the animal. An animal that is afraid may often be calmed by talking to it quietly in a firm, calm tone of voice. Shouting in a harsh tone will cause it to become even more afraid, and the same will happen if the person concerned is also afraid and displays that fear in the tone of voice. It is not what you say, but how you say it, that matters. Tone of voice is of the greatest importance throughout training.

The touch of the hand is equally vital. I have seen dogs which, through fear, would snap at the approach of my hand, relax when I got hold of them and stroked them gently. A common mistake is to clutch at the dog. This may hurt it and will almost certainly give it a feeling of being caught, which any frightened animal will instinctively fight against. Obviously, if it is afraid, you will have to hold it firmly or it will escape. Although difficult to explain on paper, there is all the difference in the world between a firm, gentle hold and gripping or clutching.

Another mistake is patting instead of stroking. Most dogs enjoy a friendly pat from their master, but the puppy we are discussing has no master, and no confidence will be transmitted through the patting of a stranger. Stroke gently but firmly, especially on the head, the cheeks or behind the ears, and the puppy should gradually relax. Dogs hate being patted on the head, although many are forced to tolerate it.

In approaching an animal it is astonishing how people manage to invent so many methods which they should *never* use. The commonest mistakes arise from the facts that all dogs hate being stared at, and that they view with suspicion anyone who falters in his or her approach. The answer is simple – do not stare and do not hesitate. At the same time do not rush towards a strange dog.

HOUSE TRAINING

If you intend bringing it up indoors the first thing you will want to teach a puppy is to be clean in the house. This should be started right away. If it can be persuaded to develop the good habit of going out to relieve itself it should prove no further trouble, whereas, if it develops the bad habit of using the best carpet for the same purpose, it may be extremely difficult to cure.

It seems to me that there are two standard methods of 'house training', one being to rub the poor little mite's nose in what it has done, the other to smack it with a folded newspaper. It is hard to say which is the more horrible. Young animals cannot go for any great length of time without emptying both bowels and bladder. Yet mothers will wrap their own child up in a nappy, and wallop a puppy for behaving like a baby.

Success in house training depends on attention to the following: (1) Practically any puppy, brought up under clean conditions, *wants* to be clean and it is up to you to allow it to develop this instinct; (2) No eight-week-old puppy can go more than a few hours without relieving itself; (3) Regularity in feeding and exercising produces regularity in the working of the puppy's inside; (4) Attentiveness and powers of observation on the part of the owner are essential.

As soon as a puppy settles down, it usually decides on a certain spot where it always goes to relieve itself. It is up to you to see that this is outside. A clean puppy, when it wants to relieve itself, will give some consideration to the place it uses. Although it may not 'ask' to go out by whimpering, the symptoms that it wants to do so should be obvious. When these symptoms appear, take the puppy out, quietly and without fuss, and wait till he does what he should do. Note that I say *take* him out. All too often, the puppy is pushed out, the door shut, and it sits on the step until the door opens again. It then comes back in and does what it wants to do where it originally intended doing it.

Many house-training problems arise from the fact that the puppy goes out for a walk, comes in, and straight away does indoors what it should have done outside. If it always goes to the same spot, the instinct to be clean is strong; the trouble is it has got things the wrong way round. Several things can be done without resorting to punishment. You can take it out for its usual walk, bring it in and, before it has had time to do anything, take it out again. If it always uses the same mat, remove the mat. You can alter feeding or exercising times – anything which breaks the routine may break this habit.

A useful tip to make a puppy empty his bowels immediately is to insert a matchstick in the anus, in exactly the same way as you would insert a thermometer; let the puppy go and he should immediately get rid of it and everything else. Very soon it should develop the habit of relieving itself immediately it is let out. This method is most helpful with puppies that persist in waiting until they are indoors.

Here we come to another advantage of a playpen. If, when you leave the puppy for the night or at any other time, you put some newspaper down in his pen, it is easy to pick up if he happens to soil it. I am not suggesting that you encourage the puppy to regard his pen as a lavatory, although that is preferable to his using the whole house as one! If he gets into the habit of relieving himself on the newspaper, he should, when there is none there, look for it, giving you the opportunity to pick him up and take him out.

Perhaps the most important point of all in house training is regular feeding and exercising. It is generally agreed that little and often should be the maxim in feeding puppies, with as short a gap as possible between the last feed at night and the first in the morning. In theory that is all very well but, in practice, it causes a great deal of trouble. To give a puppy a large bowl of milk last thing at night and be surprised to find an equally large pool on the floor in the morning is surely ridiculous.

At one time my puppies were fed on orthodox lines, six times a day, then five, then four and so on, and everyone commented on their wonderful condition. They now have only three meals a day by the time they are eight weeks, with the last at 7 p.m. and the first at 7 a.m., and people still comment on their wonderful condition.

It is seldom indeed that a puppy of about six months old is not perfectly clean in its kennel and run.

Naturally, the strength of the instinct to be clean varies

considerably and may, in rare cases, be absent altogether. Quite often it is considerably weakened by a puppy having been reared under filthy conditions. If you have been unfortunate enough to acquire such a puppy, the above methods will not work and you will have to resort to corrective methods. This should never be done until the puppy has gained confidence in you, and until you are quite sure that the above method is having no effect.

We must now return to association of ideas, our object being to make the puppy associate mistakes indoors with an unpleasant experience. The success of all corrective training, whether applied to the dirty puppy who knows no better or to a deliberately disobedient adult, lies in catching him in the act. I have, on occasions, waited for days and weeks for an opportunity to catch a dog as he was actually doing what he should not do (e.g. chasing cycles or rabbits), but, by making full use of that opportunity, have often cured the dog in *one* lesson.

That is not likely to happen with a young puppy, though it sometimes does. The difficulty arises from the fact that the type of puppy I am discussing just does what he wants to do wherever he happens to be. Unlike the puppy whose instinct is to be clean, this puppy will not trouble to look for any particular spot and, therefore, gives no warning. The only thing to do is to keep a constant eye on him, and, when he does squat, grab him by the scruff, show him what he has done, give him a shake and put him out. If you catch him in time he will not have done anything and will probably do it outside, when you should praise him. Do this once or twice and he should soon become apprehensive when he feels uncomfortable. If you are observant, this will be obvious and you can *take* him out. From then on you can proceed as already advised.

COMING WHEN CALLED

Two things you should teach your puppy right from the start – its name and to come when it hears it followed by 'here' or whatever command you intend using. Despite the thought and argument that may have gone into the choice of a name, to the puppy itself that name is merely a sound. Now any animal, wild or tame, will go to or run from a particular sound which it associates with pleasure or displeasure as the case may be. Cattle and sheep which are in the habit of having their food carted to them by tractor will run to the sound of the engine, which cannot be due to any instinct for there is nothing natural about a tractor. They merely associate a certain sound with food.

As soon as they start feeding (at three to four weeks) puppies will come tumbling out of bed in response to 'Puppy-puppy!', a whistle, the rattle of a dish, or whatever sound they associate with food. They will respond to several sounds, but as your puppy will have to learn to differentiate between sounds in the form of words of command, these should, from the start, be as few and simple as possible. Make up your mind, therefore, what you are going to call him and call him only by that name. You must use that nice, friendly tone of voice to which I have already referred (whether you feel like it or not!), and when he does come you must praise and reward him. At this stage food is the best reward.

Puppies often become 'artful', refusing to be caught when you want to tie or shut them up. This is almost invariably the fault of the owner. Many appear to give their pups well-planned lessons in not coming when called! Imagine, therefore, a mischievous puppy called in a friendly tone. He comes towards you, then halts, squats on his elbows, eyes beaming and tail wagging. Make the slightest move towards him and he's off! Run after him and that is terrific fun! Just what he wants, in fact. Ignore him, however, by walking in the opposite direction and that is no fun at all. The chances are that he will come trotting after you – rather disappointed, perhaps – but that is much better from your point of view.

You cannot catch a puppy to put a lead on if you just keep walking in the opposite direction, and we will imagine that you have one which has just started to develop this awkward habit. You call him, he comes so far but no farther, and seems to say, 'Come on, chase me!' *Never* chase a puppy. You can't catch it, anyhow! Instead, sit or squat on your heels and *ignore* the puppy. Puppies appear unable to resist the temptation to investigate, and, if you do that the first time your puppy halts and wants to play, it is almost certain that it will come and put its paws on your knee, when, of course, it should be well rewarded.

If, however, it has had one or two games with you it is unlikely that it will come right up. It will come closer but not near enough for you to touch him. Without looking him in the face, and without moving your body, extend one hand towards him, at the same time talking to him in a friendly, persuasive tone. I realize how difficult that is if you have more important things to do, but it will be quicker in the end and save endless trouble in the future. If you do not move towards him, he will in time move towards you. As he does so, put some enthusiasm into your voice to try to encourage him further, but do not move anything except your fingers.

Without training, most dogs will come up to a friendly hand extended towards them with fingers moving. Eventually the puppy should come up to sniff or lick your fingers, when the natural temptation is to grab the little blighter when he is within range! If you succeed in catching him this time, which is unlikely, you certainly won't catch him next time. Instead, withdraw your hand gently, at the same time coaxing him to come with it. As he comes nearer you will be able to stroke his head and cheeks gently. Gradually draw the hand away from the puppy and he should follow it, eventually ending up close beside you. Then, and not until then, reward him by making a great fuss of him and giving him some food, and let him go.

'Let him go?' you protest. A natural reaction, but remember that, unless you get him out of this habit, it will almost certainly get worse and you will have even more trouble every time you want to catch your puppy. So let him go, give him a few minutes' freedom, call his name, squat down, as before, and you may be surprised to find that he comes straight up and sticks his nose into the palm of your hand. He may not do this first time but he should be better each time you repeat the performance. Very soon he should, on hearing his name, come straight up to you in anticipation of the reward which you must never forget to give him.

If, at any time, your puppy is coming to you of his own accord, call him by name and praise him well when he reaches you, but *never* call him when you know perfectly well that he will not answer anyhow. Every time you give a command which is obeyed you have taken a step up the training ladder; every time it is disobeyed you have taken a step down. 'But,' you say, 'have I to stand like a fool watching my pup running away and do nothing about it?' The answer is that, unless you are an Olympic sprinter, there is nothing you *can* do. You can follow him, of course, and if you are lucky, he may decide to return to you.

Whatever you do, don't shout the puppy's name. As we proceed with training the uses to which the name may be put increase. At this stage it must only be a friendly sound in response to which the puppy comes to you in anticipation of some reward. It should be used *for no other purpose*.

A dog's name or a command repeated over and over again, without association with anything in particular, becomes a sound that means nothing. If my dogs are all playing together and I call any one of them, that *one* dog comes to me but the others pay no attention. That is not due to all the other dogs saying to themselves 'That's old So-and-So being called, I'm not wanted,' it is because only *one* dog has been taught to come to me in response to that name. All the other names mean nothing, they are sounds with no association, and, where a lot of dogs are exercised and worked together, there are so many of these names being constantly repeated that the dogs to which they do not belong ignore them completely. In exactly the same way, many dogs are taught to ignore their *own* names by owners who keep repeating them for no reason at all.

While teaching your pup his name, you should also teach him to associate a certain sound with praise, so that eventually you will be able to praise him at a distance. Every time the puppy comes when you call him make a great fuss of him. As you do so, repeat in a *very* friendly tone 'Good dog,' 'Good boy,' 'Good girl,' or whatever you fancy. Very soon your pup, on hearing this sound, will realize that he has done the right thing. By associating a particular word of praise with reward a trained dog will, on hearing it, show obvious pleasure. It is then possible to reward him by voice only. The value of this you will see as we go along.

You may be saying to yourself, 'That's all very well, but I have done all that and, when I call my pup, he just continues to do what he was doing.' That is the independent 'pack leader' type, and as training depends on balance between correction and reward, when reward fails correction must be applied. Requests are replaced by orders. The difficulty here is the distance between yourself and the pup.

This can be got over by either of two methods, the safer, although slower, being the check-cord method used generally by gun-dog trainers. Assuming that your pup is accustomed to collar and lead, put him on a collar to which is attached about thirty feet of light cord strong enough, but not unnecessarily heavy, for a pup his size. Nylon cord is ideal. Now take him to a place where the cord will not become entangled, and let him run 'free' until he is about twenty to twenty-five feet away. When you feel certain that he will *not* pay attention, call him in a friendly tone. Nothing happens! Now change your tone and repeat his name and command 'here' as an *order*.

Still nothing happens! Do not wait until it does – follow the command with a sharp jerk on the check cord, remembering that you cannot jerk a cord that is already tight, nor can you time a jerk accurately if you first of all have to coil up a lot of loose cord. This will attract the pup's attention, when you should repeat his name in a friendly tone. There is a possibility that he will now come right up to you. If he stops, repeat the command in a firm tone and, if necessary, follow it with another jerk, again followed by coaxing in a friendly tone. And do not forget to praise well when the pup does eventually reach you.

The object is to get the pup to associate this firm command with correction (a jerk), in anticipation of which he will soon move towards you *before* you apply the jerk. You must, therefore, give him the opportunity to do so. Give the command, wait a second or two; if it is ignored jerk the cord, but if it is answered change the tone of voice completely and call the pup in a friendly tone. Don't wait all day – a dog should do what it is told *when* it is told. Remember that the check cord is merely a means of communication, enabling you to apply correction at a distance. Simply to drag a pup to you on the end of a cord will do more harm than good.

The other method of correction at a distance (one I usually use myself) requires much more skill and is more likely to go wrong. In it I bridge the gap between myself and pup without any 'communication cord' by throwing something at him. Imagine that your pup is investigating a most interesting smell about five or six feet from you. You call him in a friendly tone, knowing quite well that nothing will happen. You repeat the command in a firm tone but this time, instead of a jerk on the collar, something descends upon him from heaven. What this is depends on circumstances. A handful of loose earth gives marvellous results and its spreading effect is helpful to those whose aim is not all it might be! This will give the pup a fright and he will run for protection – to you – when you must praise and console him. Next time he hears the command in the same tone he should, in anticipation of a similar experience, rush to you in order to escape.

Although this is the best method I know of getting a dog to come when called, it is much more liable to go wrong than the check-cord method, when it may well have the opposite to the desired effect. For instance, the pup must not realize that you threw anything at him or he will, quite naturally, run away from you, not to you. It is possible, however, that, as you are about to throw something, he looks round – he may even decide to come to you after all. It is not everyone who has the alertness of mind necessary to change their tactics instantly and completely to reward instead of correction.

Neither of the above methods of correction is dependent on the infliction of pain. As I have said, dogs are particularly insensitive to physical pain, and shock tactics working on the mind provide quicker and more lasting results. They should, however, be used only on dogs with stable minds. Ideal for the strong-willed pup, they are *quite unsuitable* for a very young or rather timid one.

CHEWING

One habit of which practically all puppies reared indoors have
to be broken is indiscriminate chewing. As with all bad habits,
the first step to either cure or prevention is to find out why the
dog does it. This is easy in the case of chewing, which puppies
do firstly in response to the instinct to catch hold and worry
something, and secondly because it is Nature's way of getting
rid of the baby milk teeth. For that reason this habit usually
develops during teething, and it is for the same reason that
you should not try to prevent chewing but merely to prevent
chewing the house to pieces.

Prevention is the best line of attack. When you cannot keep
an eye on him, put your puppy in his pen or kennel and give
him something to chew. A good, big raw bone is best. Having
given him something to chew, you must now teach him that
everything else in the house is taboo. You must be consistent;
it is no use saying today 'Never mind, that's an old hat anyhow,' then beat the poor little brat tomorrow because he treats
your best one in the same way.

The lessons in house manners which your pup is going to
have should be helpful in his further education. He should
learn to associate chewing anything, other than his toy, with
correction, and to associate a certain sound with that correction – to learn that when you say 'No' you *mean* 'No'.

As soon as he begins to feel at home, practically any pup
will find it great fun to seize anything that affords a good grip.
Do not try to pull it away – that is just what the pup wants.
Take hold of the pup (not the object) firmly but gently by the
scruff in one hand and in a harsh tone say 'No,' at the same
time giving him a slight shake (a tightening of the grip rather
than a shaking proper). This is sufficient to make a sensitive
pup release its grip, but some tough ones will grip all the
tighter in case they lose the object. In that case repeat the
command 'No' just as before (don't shout), following it with a
firmer shake, at the same time giving the pup a light tap on the
nose with the other hand – with one finger to start with. If that
has no effect repeat the process, each time shaking and tapping harder until the pup releases its hold.

You may think that will make the pup afraid of your hand

and, unless you are careful, it will. There is no need for that to happen, however, if *immediately* the pup releases his hold you change your whole tactics completely. Keep hold of him, but change the tone of voice from one of scolding to one of praise and, instead of 'No', tell him he is a 'Good dog' (and say it as though you mean it). Most important of all, the hand which corrected him for chewing by tapping his nose must now reward him for letting go by stroking and fondling, the object being to get it firmly established in the pup's mind that, so long as he does right, you will *never* harm him.

By the time you have done this several times you should find that, when you say 'No' in a scolding tone, the pup will let go, or at least falter in his game. If he lets go you must tell him he is a 'good boy' in a praising tone of voice. If he only falters and you praise him *at the right psychological moment* he will let go altogether. If you miss the right moment you may be praising him, not for letting go, but for holding on.

Gradually you should reach the stage where, if the pup decides to chew the carpet at the other side of the room, all you have to do is say 'No' and he will stop. Likewise you can praise him for obeying. That may be the first step towards your being able to apply correction and reward by word of command only. In this way you can tell him whether he is doing the right or wrong thing. Just as useful in teaching him to run to left or right of a flock of sheep as in teaching him to leave the carpet alone.

It does not follow that, because a pup knows what it should and should not do, it will never err. After all, humans often do what they know is wrong, so why shouldn't dogs? Until they accept the fact that life is very pleasant if they keep out of mischief but equally unpleasant if they do not, most pups, especially clever ones, will 'try it on' every now and again.

Once a dog knows that it is doing wrong, but only when you are quite sure that it knows, you can employ different tactics. In the case of a pup chewing something which he knows perfectly well he should not chew, shock tactics are probably best. You can dispense with catching the pup first and simply give him a sudden and *unexpected* rap on the nose with the fingers. As mentioned earlier, these tactics should be employed *only* on bold pups, and if perchance you frighten the pup so that he runs away from you, you *must* make friends with him again there and then.

COLLAR AND LEAD

Although there are many good working sheepdogs which have never had a collar on their necks, every dog should be taught to lead properly. All too often it becomes suddenly necessary to put a dog on a lead when it is most inconvenient to spend time getting him accustomed to it. This may happen when there are other complications (you may, for instance, want to take the dog in traffic), resulting in a completely bewildered, if not terrified, dog. Time spent at home quietly teaching a dog to go on a lead may later prove to be time well spent. A pup can be taught to go on a lead at any age providing it knows you, likes you *and will follow you without one.* I never put a lead on a young puppy until it will follow me on its own.

Actually no teaching is necessary. All you have to do is let the puppy find out for himself that, when a collar and lead are put on, there is no escape, no matter how much he struggles or how much noise he makes. Let him also find out for himself that to pull on the lead, either forwards or backwards, is decidedly unpleasant, whereas to walk on a loose lead causes no discomfort. Don't just drag the poor little devil behind you. Coax him by talking in a nice friendly tone and fondling him whenever he does move with you.

One never knows when it may be necessary to tie a dog up and it is a good idea to familiarize him with the situation before the necessity arises. But don't tie a dog to the bottom of a fence on a lead long enough to reach the top or he may jump over and hang himself.

Remember that it is an offence to allow a dog to be in a public place without a collar with an identification tag.

8
Training the Young Dog

Initial Training – Starting to 'Run' – 'Wearing'
Running to Both Hands – Forcing and Gripping

INITIAL TRAINING

MOST sheepdog trainers agree that a dog should have a good deal of elementary training before he ever sees a sheep. There are differences of opinion, however, as to how much is necessary. Some teach the dog practically all the movements used in working, often using artificial devices to get him to do it. Amongst these trainers are some of the most successful of trial men. However, I do not use these methods as I find they bore me even more than the dog. I also know trial men, just as successful as those mentioned above, who work on the same principle as myself.

Before I start a dog working on stock it must drop instantly both to 'Down' and a whistle. And it must come right up to me when I call its name followed by 'Here'. I also like it to respond to correction and reward by word and tone of voice. A dog that responds to 'Good dog' and 'No' spoken in contrasting tones is a very different proposition from one that has no idea what you mean. And the 'civilized' dog, brought up as a friend, will understand what you mean much better than one tied up outside.

When starting any new exercise always choose a place as free as possible from distractions. Bear in mind that what distracts one dog may not disturb another. Some dogs, if brought up to it, find nothing distracting in a room full of rowdy children, with the radio blaring full blast. (Not that I suggest training under these conditions!) Practically any dog in the middle of an open field will be distracted by an object moving in the distance. The stiller and quieter the atmosphere the more is he likely to be distracted. What I am trying to emphasize is that a quiet spot may not be free from distractions. Anyone can keep a dog with him in body but it is the mind we want to work on. It is much easier to keep the mind with you in surroundings to which the dog is thoroughly familiar.

If your dog will come when you call him and will follow you on or off a lead you can teach him to lie down. To start with put a lead on, take it in your right hand and put your left foot on it so that it can run beneath the instep. Next pull up with the right hand and push down with the left across the dog's shoulders. As you do, give the command 'Down' in a harsh tone. It is quite often possible to get a dog to lie down through fear by simply threatening him in a harsh tone. I am sorry to say that many sheepdogs are taught in this way, extra incentive coming from the brandishing of a stick! Although you do not want to do that, you should make some use of the dog's instinct to crouch down in response to a harsh tone. He will soon learn that, when he does, nothing unpleasant happens.

Having made the dog lie down by sheer force, sometimes quite a struggle, wait until he relaxes for just a second, then praise him gently. Some trainers disagree about this, maintaining that to praise a dog when he lies down will immediately make him get up. That is true, but if you promptly change your tone to a harsh 'Down' and put him back he will soon realize that he is being rewarded for staying down. By using *both* correction and reward, the dog will go down much more willingly and more pleasantly than by using correction alone.

When the dog will lie down, the next thing is to get him to stay there. This can be done by putting your foot on the lead close to his collar and pressing it on the ground. Stand up straight, and if he attempts to get up, give him the command 'Down', if necessary pushing him down again. As he cannot get right up, he will soon learn that the only comfortable position is lying quietly by your foot.

You now have the dog lying still beside you but have not yet taught him to lie down. All you have done is to accustom him to being pushed into position and to stay there because he cannot get up. When he does that without protest you have got over the first obstacle; the next is to get him to do it on command. When you give the command, wait a second or two, then, if nothing happens, give the lead in the right hand a slight tug and give a push on top of the shoulders – but do not keep up a continuous pressure. If he shows the slightest response, praise him well *as he responds*, and if he falters give him another tug or push. The object is to get the dog to lie down on command in anticipation of the correction which will follow if the command is ignored. That is the second obstacle over.

Now we come to the third obstacle, getting the dog to stay in position while you move away. The secret of success in this exercise is to make haste slowly. Before you try to get a dog to stay down while you walk away, make sure that he will stay, *for a considerably longer period*, beside you. Time can be saved in teaching this if you take the dog in the house with you. If, when you are having a meal or reading the papers, you keep him by you with your foot on the lead, he will very soon get into the habit of lying quietly beside you.

Once he does that, but not until, you can start teaching him to stay down while you move away. Begin with the dog on a lead, tell him to 'Down' and stand facing him. Now move a step backwards and if he shows signs of getting up, order him to 'Down' *before he does it*. If he stays, move back to him, praise him and move away again.

Now move to one side, then to the other, lead still in hand, and do it all quietly and smoothly. If he stays, praise him again and, still holding the lead, try to move right round the back of him. Here you may find that he wants to turn to face you and you must correct this immediately. Keep the lead on until he will let you walk round the back of him and step over him from either side. In case you are thinking that you have no time to go through all this, I might mention that I can usually bring a dog I have known for about a week to this stage in about ten to fifteen minutes – when I usually end the first lesson.

The next stage is surreptitiously to drop the lead beside the dog, hoping he does not realize that you are no longer holding it. This will enable you to move farther away and gradually increase the distance in all directions. If the dog gets up and comes towards you, don't just shout 'Down' and be content if he obeys. This is one of the few occasions in training when you go towards your dog. Do this quickly and quietly, take him by the collar and, with a firm 'No', take him back to the exact spot from which he got up and order him to 'Down' and 'Stay'. Once he will stay in a familiar place you can try him where there are distractions.

If he shows the slightest tendency to get up before you call, put him down again firmly, and finish the lesson by going back to praise him. And do not forget that, if you are just going to tell him to get up, and he gets up before you have done so (even a split second before), he has disobeyed you. Put him down again, just for a second, and make him stay there until *you say* he can get up.

STARTING TO 'RUN'

Once you have a pup that comes up to you when you call him, lies down where and when you tell him and is generally responsive to you, all you have to do is wait until he starts to 'run'. This waiting often calls for far more patience than that required to train the dog once he has started. Of course, some pups will want to work when they are quite small, but they should not be allowed to do so. There is something quite fascinating in a little puppy creeping about after a hen but, unless it is checked, there is a risk of bad habits forming. 'Checked' is not really the best word to use. As I said earlier on, the herding instinct is now in the seedling stage – the stage where it can be very easily damaged. Actually, if a pup starts working the hens at eight weeks old, the chances are that he has an exceptionally strong herding instinct (probably far too strong) which will take some killing. Still, there is no point in taking risks which are quite unnecessary. Do not, therefore, scold the pup for working the hens.

It will do no harm to let him 'wear' one occasionally but do not leave him running around on his own. The bad habits a pup *can* get into, by being allowed to run about amongst stock on his own would fill a book. As a keen pup will find poultry, sheep or even pigs much more attractive than you, the best thing to do when they are around is pick him up (if he is small enough) or slip a lead on him. You *could* throw something at him and teach him to come when called (as described on page 120), *but* he might associate this correction with not coming when he is called, and he might associate it with working. There is just as much chance of curing him of working as there is of curing him of not coming when called. Life would indeed be dull if we never took any risks but, as I said, there is no point in taking unnecessary ones.

The age at which a pup may be allowed to work depends on how well developed he is (mentally and physically) and the class of stock he is going to work. No pup should ever be allowed to try to work stock that can run faster than he can. Unless he has a quite abnormal instinct to run wide, it is almost certain that his efforts to get ahead of the stock will be unsuccessful. His desire to do so, however, will not weaken and, before you realize what is happening, you have a dog which chases stock away from you instead of bringing them to you.

Once a dog starts this, it is extremely difficult to get him to 'wear'. Even if you do, it is unlikely that he will be reliable for a long time. If he tires or if the sheep are some distance away, there is always the risk that he will take them away instead of bringing them to you. A good driving dog can be very useful but only if you can rely on him to 'wear' when you want him to. After all, nearly any terrier or mongrel can easily be taught to chase sheep away from you!

Having trained my first sheepdog I was very pleased when a neighbour gave me a pup which I could start from scratch. My youthful enthusiasm, however, was soon to overpower my patience and I started him as soon as he would work, at about four and a half months. When Garry was about a year old I took him on the road with a drove of about two hundred sheep, far too many for a dog of his age. With a young dog I usually walk in front and let the dog 'wear' the sheep behind but I thought he was past that stage. The sheep set off at a rattling pace, and by the time they reached the end of the farm road were some distance ahead of me. But on reaching the end they turned left instead of right and I sent Garry to head them. They were round the corner out of my sight before he reached them but, confident that he would bring them back, I did not hurry. Reaching the end of the road I saw the sheep racing as fast as they could *in the wrong direction*, the dog following behind!

Of course my dogs did not get away with that sort of thing. I just had to get Garry past those sheep. Unfortunately Garry had other ideas! He had already run nearly a mile and a dog has to really exert himself to get past 200 sheep on a narrow country lane. And Garry did not believe in exerting himself – not that much anyhow! And I didn't believe in giving in to a dog. So I kept on trying to get him to go ahead while the sheep proceeded farther and farther from home. Eventually it became obvious that this dog would not pass these sheep so, taking him with me, I 'ran wide' through some fields and came out on the road ahead of the flock. I then drove them home, exchanged Garry for Floss and made the journey without further trouble. Two things I decided then – to sell this young dog at the first opportunity, and to be more careful how I started pups in future.

Eight or nine months is usually quite young enough to start a pup on sheep. If he is really keen he can be started on poultry a good deal younger than that. I start all my pups on ducks and quite a number of specialist sheepdog trainers do the same. They are easy to move, yet nothing like as fast as sheep. As it is easy for the dog to keep them under control so it is easier for you to control the dog. This is very helpful when you are teaching the pup the different commands. For the same reason a young dog should never (if it can be avoided) be started on cattle. To turn a bullock calls for considerable physical effort and mental concentration – concentration on the bullock, not on you. If you try to tell him what to do it is doubtful if he even hears you. It is like trying to talk to some-one who is engrossed in a newspaper or, for that matter, like calling to someone who is chasing a bullock. If he does hear you, he certainly does not respond.

A dog that is required to work sheep should not be kept on ducks too long. As soon as he is strong enough he should be put on to sheep. Most shepherds work an old dog and a

young one together. This is not because the young one learns
from the old but because the young one can be used on the
easy jobs while the old one is always there to cope with any
emergencies. If, for instance, an obstinate ewe faces the pup,
the old dog can be sent in to put her in her place. The
youngster will not know how to cope with this emergency and,
if his confidence is destroyed at this stage, he may never be
able to cope with it. Not only that, if an old ewe learns that she
can boss a dog, she is going to make every effort to do so
in future.

If, therefore, you have only one dog it is most important
that you start him on sheep that will run. It may well be worth
keeping a few sheep especially for the job. This need not
create added expense to the training of your dog. Provided
you have fences to keep them in ten or a dozen ewe lambs or
gimmers of any of the hill breeds will live on the place and can
usually be sold at a profit. By adopting this course, you will be
able to do your dog a great deal of good without the risk of
doing your in-lamb ewes any harm.

At whatever age your pup starts to run he must never be
discouraged. To *prevent* a dog working is a very different thing
from correcting him for running. If your pup is old enough
and is under control you can and should encourage him as
soon as he shows the slightest inclination to run. You can
praise him by tone of voice and by word of praise such as
'Good dog' or whatever you have been using. And put some
enthusiasm into it. You are trying to stimulate the dog's
instinct to herd (no different at this stage than to chase) and his
pack instinct (you being the other member of the pack). The
less enthusiasm the dog shows the more you will have to put
into it if you want to get him going. If you chase a bunch of
sheep away from you, if one suddenly breaks away from the
others or if a cackling hen flies out of a nest, a young dog will
very often run out to head it off. I am not suggesting that any
of these are essential to get a young dog to run. I mention
them only as examples of the sort of things that tend to
start a youngster.

17 and 18 The dog which tactfully keeps its distance
will sometimes prove more successful than one that rushes in
to attempt the impossible

19 A young dog should start from behind the handler

20 Teaching a young dog to come in between the fence and ducks

21 Line held ready to check dog if he attempts to grip

22 The author with some of his dogs

23 The late Queen, Border Collie, working ducks in water

24 Australian Cattle Dog: Formakin Kulta, bred, owned and trained by the author. This red dog works sheep and cattle besides being a well-known winner in the show ring

'WEARING'

The object at this stage is simply to get the dog to run round a flock of sheep when he sees them. This he may do without any encouragement (if he is really keen you will not be able to stop him) or he may require quite a bit of persuasion. Until the dog reaches that stage there is no point in going any further. Some dogs (very few if well bred) will never run, no matter how much encouragement they receive. If you have had the misfortune to acquire one of these, get rid of him and try again. Of course all young dogs do not start to run perfectly as described in Chapter 4.

It is up to you to make the best of what you've got. I shall, therefore, try to advise you how to overcome the difficulties that are most likely to crop up. Although I must, of necessity, deal with each of these as a separate item, it is unlikely that they will occur as such. They will occur in varying degrees and combinations according to the individual dog, trainer and stock that is being worked.

To start with, never let any young dog go too far away. If he is keen to run, keep him in (on a lead is better than too much checking), manoeuvre the sheep into a bunch in the middle of the field and do not let him go until you are fairly near. Should the sheep have started to run from you before you let the dog go, so much the better.

If you have a well-bred young Border Collie he may cast out in a wide circle as described on page 53, and clap down in front of the sheep without any word of command. And it is more than likely that he will stay there. Unless you do something about it the sheep will just stand and stare at him (photograph 11). But if you can get them to move he should move too. If you come round to the left the sheep will move away to the right and the dog should head them. If you keep moving round to your left the dog should keep moving round to your right or vice versa.

The sheep are now controlled by you and the dog. You are pushing them away and the dog is keeping them in to you. If you now release your pressure by moving back, it is likely that the sheep will move away from the dog and come with you.

Keep moving back and the sheep should move with you, the dog following them. Try to keep both the sheep and the dog moving once you have them going. If they stop, you will probably have to start all over again by chasing the sheep away from you.

Remember that you are dealing with a youngster that has not the foggiest idea what you want him to do. He is responding, not to you, but to an instinct. You must, therefore, encourage him as much as you can by talking to him in encouraging tones. 'Walk on' in an exciting tone if he is slow to come on. 'Wa-a-a-alk o-on, good boy' when he does move. A sharp 'Ah' if he gets a bit too keen. It's not what you say, but how you say it, that matters. With luck, you may find that in a remarkably short time you are moving backwards, followed by the sheep, followed by the dog.

Although he is not doing it intentionally the dog is in fact 'wearing' the sheep to you (photograph 12). You may now find that you cannot get him away from the sheep. Every time they move he will head them off and he may be quite oblivious to your command 'Here', even if he answers it instantly when there are no distractions. It is likely, however, that he will 'clap' to a firm command and allow you to go up to him. If you then make a fuss of him you may be able to coax him to come away with you. If not, it is much better to put him on a lead than apply correction at this stage.

Once you have taken him away and the sheep have moved a *short* distance away, you can send him to bring them back. For this I say 'S-sh', which simply means to get ahead. It is quite astonishing how quickly a keen young dog will learn this command. In a very short time you should have a dog which, when you tell him, will run round a flock, bring them to you and keep them following when you move away. Of course, you may not be quite so lucky!

When you move backwards away from the sheep and they move away from the dog it is quite likely that he will head them and run right round between you and the flock. He will then get them milling either to the right or left and, if you are not careful, will run round and round keeping them in a bunch. Here the training you have given the dog in 'clapping' on command will be of practical value. As the dog comes round the flank drop him. If he comes round the right flank the sheep will naturally move to the left. When they start moving encourage the dog to head them on the far side instead of coming round between you and the sheep as he intended doing. When he gets far enough round drop him again and repeat the process in reverse. Soon you should have the dog moving backwards and forwards behind the sheep with you moving backwards and the sheep following you.

Some 'strong-eyed' pups, instead of casting out to gather, creep straight at the sheep and stay there 'eyeing' them. If the sheep run away, the dog will almost certainly cast out to head them. If they stand and stare at the dog, chase them away from him in an effort to get him to go.

You may not have a dog which naturally runs wide. Most 'loose-eyed' dogs, and many which are 'strong-eyed', will run straight to the sheep, come close in round them and keep them going as fast as they can. To make matters worse, some will grip the outside sheep as they go round the flank. Such dogs are usually keen enough to stand up to pretty severe correction and, if they do not get it, become the cause of so many farmers shaking their heads and saying, 'Dogs should never be allowed near lambing ewes or milking cows.'

Correction applied in the early stages will not have to be nearly so severe as will be necessary after a habit has developed. Bad habits, like weeds, are almost certain to make an appearance sooner or later. If they are firmly stamped on *as they appear* they may cause no further trouble. But if they are allowed to develop and take root they may prove impossible to eradicate.

The first essential in training the headstrong, close-run young dog is implicit obedience. The biddable youngster with a hazy idea of 'clapping' on command will learn as he goes along. The 'hard' dog should be kept away from stock until he will drop instantly *and stay there*. Once you get him behind the sheep, drop him. Coax him on, but, if he makes a rush, drop him again. It will be a long time, if ever, before you get this sort of dog working smoothly but, by continually dropping him and scolding him if he comes too near, you should be able to prevent his doing any harm. Keep a piece of hosepipe handy and, if he makes a dive in amongst the sheep or grabs hold of one, let him have it. Use a harsh word of command such as 'No' or 'A-a-ah' just before you throw. Next time the dog attempts to do the same thing, you may only have to use the same word in the same tone to get him to stop.

One often sees a young dog charging willy-nilly right in amongst a bunch of sheep. This need never happen if the dog is started close to the handler with the sheep bunched up together. He should be gradually allowed to run farther to gather sheep which are more scattered. Remember that the best way to go up a ladder is one step at a time. If it becomes a little shaky you can come down one step at a time until you get a good footing. By taking a running jump at it you may reach the top very quickly but, if you do have to come down, some of the rungs may be missing. You then end up in a horrible mess at the bottom!

You now have a dog which will run round a bunch of sheep, bring them to you and keep them up. If you do not, keep on trying until he does or you decide that he won't. Until you get to that stage there is no point in trying to go any further. In some cases you will only have to allow the dog his head and he will do it from the start, while others will require plenty of encouragement.

Whether he is allowed or whether he has to be encouraged, you should teach the dog to associate this action with a definite command, so that you can tell him to go. I use the sound 'S-sh', and most dogs broken in Scotland will answer that command. In response to it the dog should go out round anything that happens to be in sight and eventually go out to look for anything that is out of sight. Most 'strong-eyed' young dogs, once they have run out, require coaxing on by another sound like 'Come on' or 'Walk on' and some need steadying or they will come on too fast.

If you are lucky enough to have a dog that, on one command, will run out, lift and fetch sheep to you, do not worry him with extra commands. It is a practical working dog we are after, and the fewer commands he is given the more will he be encouraged to use his own brain. Dogs trained to respond to each and every command with absolute implicity often do not try to use their own initiative.

If such a dog is sent to gather sheep which are out of sight he will very probably run out till he is behind them. On hearing no further commands, however, he will leave the sheep and come back to where he can see the handler as if to ask 'What shall I do now?' But the handler, not being able to see the sheep, cannot tell him. 'S-sh' simply means to go out round the stock as quickly as possible and, unless told otherwise, bring them to you. Like anything else that dogs *want* to learn, they pick up the command to go with astonishing ease. Most keen ones will respond like a flash to 'S-sh' on the second time of hearing it. All you have to do is to give it as you allow or encourage the dog to go.

RUNNING TO BOTH HANDS

As soon as you get the dog to 'wear' with some certainty you must start him running to both hands. This is much more important in the ordinary farm dog than is generally realized. Supposing you are driving some cattle on the road and you see a garden gate open on the left. If you have a dog that runs to either hand you can send him along the left flank. He may then reach the gate *before* the first animal and should stay there until the whole herd has passed. If you did not notice the gate in time for that you should still be able to send the dog to head the first animal as it enters the gate. Cattle or sheep will rarely rush through a strange opening without the leader pausing to investigate. This pause may be for only a split second but long enough for a man with his wits about him to send a keen dog.

Supposing, however, the dog will only run to the right. First of all, you cannot place the dog in the gate on the left before the cattle reach it. And if you send the dog when they have reached the gate he will run along the right flank thereby turning them *into* the opening you want to go past. The only answer is to walk ahead of the cattle yourself and get the dog to bring them up behind. You can then stand in any openings and wait till the cattle are past. But that is not practicable where there are many openings.

Even more dangerous is the dog which runs the wrong side of a flock of sheep lying at the top of a cliff, the bank of a river or similar place. A dog running close on the inland flank of such a flock is very apt to cause the sheep next to him to push some of those on the outside over the edge. These are only two examples of the disadvantages of a dog that will only run to one hand. There are many others and it is well worth the effort to train your dog to run to both hands right from the start. If a dog develops the habit of running to one hand only, it is usually very difficult to get him to run to the other.

Many dogs – probably the majority – run more readily to one hand than to the other. If a dog runs naturally to the left, teach him to run to the right as soon as ever you can, and vice versa. The keen young dog can be made to run to both hands more quickly than one not so keen. If the latter runs to the left, and you check him in an effort to make him go to the right, he may refuse to run at all. In that case you will have to wait until he is keener.

As soon as the dog will run, make sure that he always runs either to one side or the other and never straight away from you. If the dog runs ahead of you call him back, turn yourself round so that he comes behind you and, with the right or the left hand, encourage him to go in that direction as in photographs 8 and 9. And work in a fairly confined space so that the sheep and the dog are less likely to take off into the distance with the dog crossing between you and the sheep. The fact that you are signalling him almost directly away from the sheep does not matter as they should be quite near and he knows where they are. Alternatively you make the dog lie down a short distance away, say, to your right, call him to you and, as he approaches, give him the command to go, at the same time encouraging him to go to the left with that hand.

Run the dog to one hand as much as to the other unless, as I said, he has a preference which you should try to balance by running him as much as possible to the side he dislikes. Until he is reliable on this do not let him run too far. Otherwise he may go out to one hand, cross over between the sheep and you and go round close on the *wrong* flank. This can easily become a very bad habit, difficult to cure.

By the time your dog will run to either hand, gather a flock and keep them up to you, he is of some practical value. (I doubt very much if 50 per cent of so called sheepdogs would do that!) If you want to move sheep from one place to another, even on the road, you can walk ahead and let the dog fetch them behind you. For practical work a dog should be kept at this stage for quite some time. He should, of course, still be learning. Send him gradually farther away and send him to gather sheep which are scattered over a wider area. Encourage him to keep them moving but do not let him rush them.

Do not throw away opportunities to give your dog work – practice would be a better word. Cows will come in for milking on their own, but, if you make your dog gather them and walk quietly behind, he will get the idea that they are moving for him. The hens in photograph 10 are coming to feed in any case, but the young dog is learning not to leave any stragglers behind. It is a common fallacy that only old dogs are steady. Like us, dogs become steadier as they grow older. Again, like us, they only give up their bad habits when they are too old to indulge in them! By then they are also too old to work! A young dog you can control is a far better proposition than an old one which you *hope* has steadied down.

FORCING AND GRIPPING

Naturally you do not want a dog which will only walk behind stock that are moving away from him. You want to know that, if a bullock or an old ewe faces him, he will teach it who is boss. It is a mistake to let a young dog go in to grip before he is completely under control, and before he has sufficient strength and confidence for the task. It is also a mistake to let him get into a habit of merely walking behind animals that keep moving away from him. Of course, a trial dog should never grip a sheep but we are not training for trials. If an old ewe faces a dog he has two alternatives. He can grip it by the nose, thereby reducing the chances of it facing him again, or he can move back, encouraging the ewe to become offensive. I know which I prefer and I am sure that anyone who has ever tried to work a dog that draws back will agree!

Although dogs do not (or at least should not) grip at trials, it does not follow that they *will* not grip. Trial men often go to a great deal of trouble in teaching a young dog to go in and take hold. This gives the dog confidence. Knowing that, if attacked, he will be able to retaliate *and win* he will then go right up to a stubborn sheep that faces him. Because he creeps in steadily with determination and confidence it is almost certain that the sheep will lose its nerve and move away from the dog. I have never known a dog, literally 'free from grip', which would do that.

Of course, some dogs need no encouragement in this direction – quite the reverse! The good-natured young dog, however, will be less inclined to retaliate automatically than will his quicker-tempered brother. Also, the 'hard' dog which has had a good deal of correction may be somewhat apprehensive as to what may happen to him if he grips a sheep. In either case you must encourage the dog and make him feel that he is doing the right thing. Excite him by your tone of voice, your actions and your whole attitude. Perhaps most important of all at this stage, see that he ends up top dog.

Whenever I offer advice on how to stop a dog chasing sheep I am sure to receive letters describing a '*certain* cure'. This is invariably on the principle of shutting the unfortunate dog up with an aggressive old ewe or ram and letting the latter do the training. Undoubtedly many dogs have been cured of chasing sheep by such methods, but sheepdog trainers have been known to teach dogs to *attack* by the same method. If a dog and an aggressive sheep are shut up together and left to their own devices, one of two things will happen. The sheep will teach the dog not to chase sheep or the dog will teach the sheep not to chase dogs. It all depends on what sort of stuff each is made of.

We are back again with our balance which can be tipped either way by the trainer. It is possible to teach a sheep not to chase dogs or to teach a dog not to go near sheep – *with the same sheep and the same dog*. If I go to the aid of the dog I can drive the sheep off, at the same time encouraging the dog and making him believe that *he* has driven it off. This gives him confidence and next time he will require much less help or encouragement. But, if I go to the aid of the sheep by scolding and threatening the dog, I can get quite the opposite result.

The dog will first of all hesitate, leaving an opening for the sheep to attack. He will then lose his nerve completely, resulting in a sheep that will chase the next dog it sees and a dog that will not go near a sheep – so long as he can get away, at any rate. Forced into a corner, there are remarkably few animals that will not fight for their lives and dogs are rarely the exception. A dog that is forced to retaliate by having his back pushed to the wall is likely to attack *before* he finds himself in the same position again. His fear of being cornered, however, is liable to make him want to get one in first, often losing his head and attacking viciously.

Although of more practical use than the one that runs away, I cannot recommend such a dog for general purposes. Of course, no two dogs, sheep or people are alike, so that it is impossible to divide the combined efforts of a dog, a sheep and a person into distinct categories, just like that. What I want to emphasize is that, just as your dog must never get the better of you, so must other animals never get the better of your dog. If there is any danger of the balance being tipped against him, see that you are in a position to tip it in his favour. Do not, for instance, send a young dog four or five hundred yards to fetch an aggressive old ewe with a young lamb. You will have to go and fetch her yourself in the end, so you might as well go to begin with and avoid the risk of spoiling a young dog.

As I said earlier, you can only work on what is there. If a young dog just hasn't got the guts, neither you nor anyone else will put them there. As I also mentioned, however, a seedling which can easily be killed can, with proper treatment, be encouraged to grow into a vigorous plant.

Another reason why you should always try to be near a young dog when there is a likelihood of his being faced is that he may retaliate with more enthusiasm than you expect. Do not ever let a dog develop the habit of hanging on to an animal. If a sheep faces him, let him give it a good one, but, as soon as it turns, see that he drops back *immediately*. If he does not, the sheep may either panic or be forced to attack again in self-defence.

9

Further Training

Teaching 'Sides' – Coming On – 'Lifting' From Awkward Places
Driving – Shedding – 'Holding' Single – Barking – Tracking

TEACHING 'SIDES'

WE STILL have not trained the dog to his 'sides' – to move to right or left so that he can be placed in any position you want him. The stage at which this should be done varies with the dog and the purpose for which he is being trained. For our purpose, I do not think there is any hurry. The methods of teaching it also vary. Most shepherds use hand signals to send the dog out and direct him once he is behind the sheep. Trial men do not use this method at all and the method of moving trial dogs is becoming more and more general. This is not surprising as it is much more effective.

Only four basic commands are used: to move round in a clockwise direction, to move round in an anti-clockwise direction, to come on from any direction and, of course, to stop. Hand signals are never used as, to answer them, the dog must take his eye off the sheep – a fatal mistake in trial work. The almost universal commands used are 'Come bye' to move clockwise and 'Way t'me' to move anti-clockwise.

This can best be explained by telling you how I teach a dog to respond to these commands. Some trainers teach the dog this before they start him working, making him go round themselves in both directions. As I have already mentioned, I find this type of training boring for both the dog and myself. Having a very limited number of animals to work I have evolved a method which saves them a good deal of chivvying about by over-keen and under-worked young dogs. I have also found that I attain better and quicker results.

As explained earlier, the first thing is to get the dog 'wearing' reliably. As soon as he does that, I start him working ducks in a pen as shown in photographs 13, 14, 15, 16 and 19. It is not essential to have ducks, of course – hens, sheep or even a tame rabbit will do if you have a keen young dog. Round this pen I teach the dog all his commands, and I like to have him obeying each command before he does it without the pen.

To start with I teach him to run to both hands. The young dog in photograph 19 has been brought from my right, has come round behind me and is being sent out on the left. He learned the command 'Here' as a pup, and 'S-sh' when he started running. When he is on my right, either by chance or because I have made him lie there, I call 'Jed here' and, as he reaches me, I say 'Come bye, s-sh,' at the same time encouraging him with the left hand. Sometimes it helps to touch the dog gently on the cheek in an effort to coax him out in the right direction. Of course the dog does not understand 'Come bye' but, as it is followed by 'S-sh', which he does know, he should soon pick it up. When he is behind the ducks I say 'Down', and he lies down as he was taught that before we started. I then praise him very well by tone of voice saying 'Good boy'.

Now I try to make him move again. I say 'Way t'me, s-sh', at the same time signalling with my left hand and moving slightly to the right as in photograph 13. This will tend to make the ducks move to the left, which, in turn, will make the dog want to move in the same direction to head them. *As he does* I praise well to let him know that that is what I want him to do. Then I drop him again as before.

Having got him to do that, I then repeat the whole process in the opposite direction, giving the command 'Come bye'. Once I have managed to make him move round a little way on the far side of the pen, I put him down and move over to his side so that the ducks are slightly to the dog's right and my left. I then call him to me with 'Come bye, *here*', bring him round my back and send him out to the left with 'Come bye, s-sh'. Depending on how keen your dog is – and how keen you are too – you can keep this up almost indefinitely without tiring the ducks. But do not overdo it. Always finish a lesson when the dog is still willing to go on and remember that you will start tomorrow where you finished today. If you want to start on a good note, make sure you finish on a good note.

You may be surprised at how quickly and easily you get your dog to move to the right and the left. But it does not follow that he understands his commands. Some dogs do pick up the commands this way but, as you are moving from side to side, he may only be moving instinctively to keep the ducks between you and himself. You, therefore, do not want to keep it up too long or you may accustom him to a command which means nothing. I do it chiefly to give the dog the idea of moving from side to side.

You must soon advance to the more difficult job of making him move round when you stand still. If he moves to the right the ducks will move to the left. If he is a keen 'wearing' dog, and has had the idea of 'wearing' fixed in his mind (as he should have), he will want to change direction to head the ducks. The easiest way I have found to overcome this is to put the dog on a line. Remember, in training a dog on a line, it is more a means of preventing his doing wrong than making him do what you want. It is also important that the dog should be quite happy on a collar and lead before you try to work him on a line.

With one end on the dog, the other in your hand, and no obstacles about (a round pen is better than a square one) send the dog round the pen (say to the left). But stop him before he gets to the opposite side. Having shortened up your line so that the dog can only go about a foot farther, give him the command 'Way t'me'. He is not sure of the command yet and his natural tendency will be to continue round on the circle, but the line stops him. A keen dog, finding that he cannot go that way, will very often come back round the other way in his anxiety to reach the other side. He should, of course, be praised very well if he does.

It is quite possible, however, that the young dog will be somewhat bewildered at being checked by the line. That is why you must take up the slack, as, otherwise, he may have gathered sufficient speed by the time he reaches the end of the line to check himself quite severely. This would, in fact, be correction for running, which must be avoided at all costs. If the dog is merely prevented from going any farther he will probably lie down, somewhat bewildered, but he will not have been corrected.

You can now call him back to you with 'Here' and, as he approaches, coax him with the hand and 'Way t'me, s-sh' to go the other way. If you have always made your dog go round your back when teaching him to run to both hands, you will probably find that he does the same now. Do not try to stop him. In teaching a dog any new exercise I always concentrate on getting him to do what I want, not on *how* he does it. You are trying to make the dog move round in an anti-clockwise direction. Get him to do that before worrying about whether he comes behind or in front of you.

Naturally you want him eventually to change sides at a distance without coming back to you and going out to the other hand. As soon as the dog understands 'Way t'me' you can stop him going round your back and encourage him to come round between you and the ducks. If there is difficulty in doing this, I have found that it is a good idea to stand with your back to a fence or hedge about four or five yards away from the round pen. The dog then cannot go round your back and, without having to check him at all, it is easy to make him pass in front of you.

You can only work a dog on a line in two half circles. He cannot go round and round in both directions. Once he understands the commands, however, you should have no difficulty with that. As soon as the dog is responding on the line, try him without it. If he will not respond to your commands now, it is probably because you have been dragging him about like a turnip on a string. Put the dog back on the line and try to make him respond to *your commands*, not to the line. As I said, it should be used only as a means of preventing his doing the wrong thing. *You* must make him do the right thing.

COMING ON

Teaching a dog his sides is only one of the uses I have found for the duck pen. It is the easiest way I know of training a 'sticky' dog to get up or of steadying one that runs in. At the pen I teach my dogs to come on to the ducks in a straight line at varying paces and from any direction. To start with I have the dog lying at my foot several yards from the pen on a line gathered up to about three feet. I then say, 'Queen walk on', and simply lead the dog forward a few steps (photograph 15) and say 'Down'. This I repeat several times until I reach the pen. A keen dog will, by then, be rising in response to 'Walk on', before I attempt to lead him forward. Not only do I praise him if he does – I praise and encourage him if he shows the slightest inclination to do so.

Next I try to make him walk on while I stand still and allow the line to slip through my fingers (photograph 16). As most people on farms today are more familiar with spanners than with ropes, I might mention that a line will not run smoothly through the fingers unless it is coiled up properly. I now put the dog down a short distance away from me and get him to walk on from different directions. This I continue until I feel sure that he knows that his name followed by 'Walk on' means to walk straight up to his charges.

This can now be combined with what I have already taught the dog. I send him out to one side or the other from behind my back, still on the line if he is at all erratic, but without it if biddable. Although I have been trying to train the dog to change sides between me and the ducks, I now want him to run wide, so I cast him out from behind me. When he reaches a point about a quarter-way round the circle, I stop him and tell him to walk on. I then call him back and put him out behind – to the other side where I repeat the process.

To begin with I drop the dog to stop him but, as soon as possible, try to keep him on his feet. If you are in the habit of preceding your commands with the dog's name, you will find that he hesitates when he hears it, ready to respond to the next command. You send him out, call his name, he falters and you say 'Walk on'. Once he has the idea you can improve on it with another command. Send him out, call his name followed by 'Steady there' or 'Stand' and keep him standing until you tell him to walk on. You may have to shorten up your line and go back to the dog to get the idea into his head but it is worth the effort. The dog that stays on his feet is always a much smoother worker than one which bobs up and down all over the place.

We now have a dog that will run to both hands, will change sides either on the far or the near side, will stand or lie down where told and will walk on from any position. But that will not make him a worker. It will, however, make him much easier to work. Passing exams at college does not, of necessity, make a good farmer. Providing the young man has the other essentials, however, the theory he learns will help him to understand what has to be done in the field.

The duck pen can be a great help in preventing or correcting bad habits. You can concentrate on the dog, knowing that meantime the stock is not running away. If you have a keen, excitable dog that runs in too close you can throw something at him *as* he dives in round the pen. This will actually drive the dog back from the ducks and you should get him to associate it with a command like 'Get back'. If then he shows any sign of running too close you should be able to keep him out by shouting 'Get back'. When you command this type of dog to move to the right or left it is almost certain that he will come in closer to the ducks as he moves. This can also be checked by throwing something at him as mentioned above. Work on this until you can make him move in a wide circle in either direction and only come on when told to do so.

By practically the same method a 'sticky' dog can be made to move. It always amuses me at trials to see a really 'strong-eyed' dog get stuck when a sheep faces him. The handler shouts 'Get up' or some other sharp command and the dog springs to his feet like a jack-in-the-box. I do not know how all these dogs are trained, but I have a good idea! Obviously the dog associates the sharp command with something pretty drastic, most probably something descending upon him from out of the blue.

A shepherd I know obtained a young bitch which since puppyhood had been allowed to lie watching sheep through a fence. The result was that this had become a fixed habit before she was allowed to run. Taken into a field of sheep she would simply run forward and set at them as if in a trance. All efforts to get her to run out had failed and the shepherd's patience was wearing thin. One day when she ran forward and set at a sheep he walked up behind her (she was so intent she never

knew he was there), said 'Get up out o' there' and gave her a
hefty kick on the behind. She got the fright of her life, shot off
like a bolted rabbit and ended up behind the sheep, which, of
course, was where she was wanted. That method was,
perhaps, a bit crude but it worked, and from then on if she set
on the outrun, a sharp 'Get up' would send her on. The
important point is that she associated the correction with the
command. Otherwise the shepherd would have had to kick
her every time, which would not have been a great help!

A dog that really grips, especially one that has been allowed
to get away with it, will require more severe correction than
that described round the duck pen. Also a clever dog, realizing
that he cannot get at the ducks, will not try to do so. As with all
bad habits it is necessary to make the dog do wrong if you are
to correct him when he does it. If you can create a temptation
which you can control it is much better than simply waiting till
the temptation arises.

With a gripping dog I like to take him in quite a small
enclosure with a few sheep. Put him on a line long enough to
allow reasonable freedom to the dog but not so long as to be a
nuisance to yourself. With the help of the dog push the sheep
back into the corner, as in photograph 21, then deliberately
make one or all of them break away. This will try even a steady
dog and it is almost certain that one with a tendency to grip
will try to dive in and take hold. This is your opportunity to
check him really severely with the check cord. Having done so,
the sheep will still be quite near so that you can start again.

We are dealing with a very keen dog and you want to keep
this lesson up until you get some signs of response. On no
account use the line as a means of holding the dog back from
the sheep. Keep him back by word of command and, if he
does not obey, correct him severely with the line.

'LIFTING' FROM AWKWARD PLACES

Once the dog will gather and 'wear' a bunch of sheep in the
middle of a field, you can try to make him lift them when they
are lying alongside a fence or hedge. This is something which
very few dogs will do naturally, and which should be taught as

an exercise rather than wait till it *has* to be done. The natural tendency is for the dog to take the easy way and turn the sheep into the fence. When they move along it he heads them and turns them back again. If he gets away with it a dog will soon develop the habit of keeping sheep up to a fence and will even take them away *to* it instead of from it.

I like to teach my dogs this exercise with ducks or very free-moving sheep in a small space where I can keep everything under control. In photograph 20 you will see me with a young dog working ducks in one of my puppy runs. The dog has by now learnt 'Come bye' and 'Way t'me'. I push the ducks to the fence (in this case with the help of an old dog) and give the dog the command 'Come bye'. He tries to go past but finds no room, so changes direction to head them on the other side. But I have deliberately placed myself right at the best point to stop him doing so. When he finds he cannot go that way he will almost certainly have a try at the other way. At the same time I steady him up to impress on him that, when I *say* 'Come bye', I *mean* that he has to move clockwise even if it is easier the other way. To begin with he will probably make a dive for it, as a dog does not like to feel himself jammed in a small space. Once he learns that the animals move over he will gain confidence and can be steadied down till he will move in between a flock and a fence quietly and firmly.

DRIVING

Once your dog 'wears' reliably you can teach him to come in behind the sheep and drive them forward with you. Here your command of the dog will have to check his natural instinct to run out and head the flock. To begin with you can make the dog move backwards and forwards behind in response to 'Come bye' and 'Way t'me', but he is not likely to push them up at all. Once he is doing that, however, you should find that he will come in close to push them through a gate or into pens, etc. Under those conditions it is much more difficult for him to head them, and your own efforts to get the sheep to move will excite the dog, tending to make him come in. Take care at this stage to see that he does not use his teeth too freely.

To the hill shepherd a dog that will drive away is a great help and in the driving championship at the International the dog drives fifty sheep away in a straight line for 800 yards. Although unlikely to be so useful to the general farmer there is no harm in teaching your dog to do it. Start with stock that runs freely – animals that are in the habit of moving in a certain direction are useful for this. Use the same command as you use for bringing the dog on and stop him if he tries to head them. If he moves over to the left, bring him back with 'Way t'me' till he is between you and the sheep, then tell him to 'Walk on'. Do the same thing with 'Come bye' if he runs out to the right.

Vary the tone of voice. If the dog moves to the right and you give a sharp 'Come bye' he will probably change direction and shoot off to the left, then back again with a sharp 'Way t'me'. If you give a slow 'Come bye', or even just the first half of the sound, 'Come', the dog should move more slowly, when you can give him a quiet 'Steady there – walk on – good boy'.

A keen 'wearing' dog will sometimes keep trying to head the stock and here again the best thing is to put him on a long line. Run the line out to let him follow the stock but keep it ready to stop him (not to correct him) if he tries to head them. Once they have the idea, most dogs will drive stock away so long as it moves freely but they are not all so keen on really *driving*. Whether or not you use a line, follow the dog to start with but gradually allow him to increase the distance between you.

SHEDDING

In practical work it is rarely that a dog is asked to 'shed' in the way seen at trials. But the trial dog that will shed and hold a single sheep is a very useful dog for practical work. A dog that will cut out, without disturbing the rest of the flock, a lame sheep, a ewe about to lamb or any other single sheep you want to catch is a great asset.

Start by 'shedding' two or three sheep which are easier to 'hold' than a single. Get the dog to 'wear' say ten sheep to you and let them settle down. Now, with the dog on one side and you on the other, nip in and cut out three with your stick. The dog will immediately make to head them as he has been taught to do. Stop him and call him in to you with 'Come in, *here*'. Now get his mind on the three sheep and, with his help, drive them away for a bit. Having done that, send him out to gather the other seven and run them together again. Repeat this several times but try to reduce your own effort and encourage the dog to increase his. Take the emphasis off 'here' and put it on 'Come in'. Some dogs take to this very quickly and get the idea after one or two lessons.

When your dog has got the idea you can try cutting out one sheep. If you have a well-bred dog he will put far more enthusiasm into this than he did with the three. He will now keep between this single and the others, irrespective of where you are standing, as the one sheep will keep heading towards the rest. Do not try to be too clever by trying to make the dog take the single away from the others. Be content with his holding it for a few moments, then call him off. If the sheep looks like getting the better of him call him off before it does.

A dog that will 'shed' a single in that way will find it easy to cut out one that is lying away from the rest of the flock. There is a silly idea in some districts that a dog should not be taken amongst lambing ewes. Far from disturbing them, a dog that will quietly 'shed' a ewe that is about to lamb will save the rest of the flock a vast amount of chivvying about.

Do not start a dog 'shedding' before he is 'wearing' properly. Some dogs become very keen on it and sometimes, instead of gathering the flock, will cut out one and 'wear' it just for fun. This can become a very bad habit and should be guarded against.

'HOLDING' SINGLE

At this stage it is also a good idea to give a dog a few lessons in helping you to catch a single. Most dogs take to this very easily, thanks to the pack and hunting instinct. Man and dog become a pack out to catch their prey. Provided *you* are the leader and the dog is not afraid of you, little trouble should be experienced. On a hill it is often necessary to catch a single sheep in the open but in fields it is usually easier to work it to a fence.

If you have a keen dog that has been checked for running in, he may be somewhat wary about coming in steadily as close as you want him. You will have to encourage him and make him understand that he can come in close *so long as you say so.* Once he does realize that, you will have no trouble with this sort of dog. As you reach out with your crook many dogs will rush in to grab hold – to help the pack leader to catch the quarry. This rarely does any harm and is often a tremendous help.

Old Judy would catch a hen or a duck anywhere on command 'Catch it', and would hold it without leaving a mark or pulling out a feather. Floss, my first Collie, would do the same, but I have had others which could not be allowed to attempt it. It is easy to teach a dog to grab hold but very difficult, if not impossible, to teach him just how hard to hold. Some will hold without really biting at all while others just bite.

Not all dogs go in to help when the handler dives forward with his crook. Just at the time their co-operation is most wanted they draw back and let the sheep run away. These are almost invariably dogs that have been swiped at with a stick. With a change of handlers such dogs will sometimes regain confidence, but will rarely go near the stick that has hit them. And who can blame them?

BARKING

A question I am often asked is how to make a dog bark. The problem usually concerns cattle dogs which heel, hock or swing on the tail but which cannot be persuaded to keep back and bark. The real answer is to acquire a dog that does. But, if you have an otherwise good dog, it may be worth the effort of teaching him to 'speak' on command. Many people imagine that because Border Collies are quite silent in their work they are naturally quiet dogs. That is not so and, from experience, I have found them easier than most other breeds to teach to bark. Some will bark naturally at cattle, at sheep when forcing or at a sheep that faces them. Such dogs present no problem.

The reason few 'strong-eyed' dogs will bark in their work is because, as I explained earlier, the 'strong eye' is derived from the wild dog's instinct to stalk its prey. If it barked it would obviously defeat its whole purpose. The 'strong-eyed', silent dog must therefore be taught to bark when there is no stock near. Learning to speak on command is an exercise which some dogs learn quite easily, while others are almost impossible to teach. Incidentally it is also an exercise over which some trainers rarely find any difficulty, while others rarely succeed.

The first thing in teaching a dog to bark is to find out what makes him bark. For instance a dog in the house may bark at a knock at the door. When he does, do not tell him to 'Shut up', as so often happens. Excite him and tell him to 'Speak'. When he does, praise him well. If he keeps on barking tell him firmly 'Quiet', and again praise him well when he stops barking. As in all training keep the object of the exercise uppermost in your mind. It is important that you get him to speak but at this stage less important that he should stop when you tell him. Remember too that the dog cannot now be rewarded by being allowed to work. Like the obedience or the circus trainer *you* will now have to provide the reward.

Which brings us to another idea you may try. Greedy dogs will often bark at the sight of their food, especially if they are tied up and it is held just out of reach. Excite the dog and tell him to speak and as soon as he makes the slightest attempt give him the food. Modify this to tit-bits you can carry in your pocket and *gradually* get the dog to speak more and more before rewarding him. This is the method by which most 'counting' dogs on the stage are taught and I usually try it first. But if at first you don't succeed, try anything that will make the dog bark.

Once the dog will bark on command, you can try making him bark, again on command, at stock. It is a good idea to keep him on a line to start with so that he cannot rush in. In fact a dog that wants to rush in will often bark in exasperation if he is held back on a line. Once he acquires the habit of it, he will probably bark of his own accord but, if he wants to go in and grip, you should be able to check him and tell him to

bark. 'Hunters' in Scotland are trained to go out for long distances, then stop and bark on command to start the sheep
moving. The usual command is 'Ho ho', which, of course, is
the sort of noise most people make when forcing sheep or cattle
into pens, etc. It is, therefore, probably better than the command 'Speak'.

TRACKING

One of the easiest things to teach any clever dog is to track,
and it is much more useful in a farm dog than is generally
realized.

Tracking a human is something a dog may never be asked
to do but it is almost certain that a dog which will track an
animal will be useful one day. I have a little bitch now, and
have had several in the past, which, if carefully started at a gap
through which a sheep has escaped, would track that sheep
and bring it back through the same gap, without my moving
from the spot.

All that is necessary to teach a dog that sort of tracking is to
organize a few 'escapes'. For instance, you can move two or
three sheep into a field where they are easily hidden from
view. Let the dog help you put them through the gate, shut it
and take him away. When the sheep have gone out of sight of
the gate take the dog back. If you give the command 'S-sh' the
dog will almost certainly cast around the field especially if it is
one he knows. But if you hold him on a line and steady him
with a 'S-sh, s-e-e-k', he will probably start pulling in the direction of the sheep. It is a good idea to know the direction that
the sheep went and it is essential in the early stages that the
dog should know there *are* sheep in the field.

Follow the dog towards the sheep and when he can see
them (being taller you will probably see them first) loose him
and send him to gather as usual. If the sheep are upwind of
the dog he will probably pick them up with his nose long
before he reaches them. This will make him much less
inclined to follow the track and in training it is best to arrange
tracks downwind of the dog. For those unfamiliar with the
terms, tracking upwind means that the wind is blowing from

the quarry towards the dog, and downwind is the reverse. It should be remembered that if there is a fairly strong cross wind a dog may track several yards to the side of the track. This is a big subject and quite fascinating but I must not spend too much time on it here.

Returning to the dog we are trying to teach, the object is to make the dog realize that if he follows the track he will eventually find the sheep. How you get him to do that does not really matter and some dogs will do it without any encouragement at all. Once a dog has grasped the idea of following a track I have found that no further training is necessary. Which is not surprising when one remembers that this is the method by which the wild dog very often catches his dinner. Much training and a great deal of practice are, of course, necessary if a dog is to be able to follow a cold line over foul ground. To follow an animal which has recently escaped is much easier but nevertheless extremely useful on occasion. Even if the dog merely indicates the direction in which the escaped animal has gone it may be of the greatest help.

10

Other Jobs and Problems

THERE are many jobs on the farm which any good Collie would be delighted to take on but which the owner rarely allows him to do. This is due either to the idea (usually erroneous) that it will spoil him for his legitimate work, or ignorance of how much any intelligent dog can be taught. We therefore find farmers keeping a Terrier to kill rats, a gun dog for shooting and a sheepdog for cattle and sheep. None of them has anything like enough work and if the three go off together trouble will not be far away. The Terrier or the gun dog is not likely to be much use with the sheep but many Border Collies will kill rats as well as any Terrier and work to the gun as well as a Spaniel.

Many shepherds will not allow their dogs to kill rats in the belief that it will tend to make them grip. This is quite a natural idea but I have been unable to find any evidence to prove it correct. My old bitch, Judy, could kill rats with any Terrier but, as I said previously, she can be allowed to catch a hen or duck without fear of her hurting it. I have taught several Border Collies to do 'man work', and although they would bite viciously at a padded arm they never showed any more tendency to grip than previously. Of one thing I am certain – assisting his master to kill rats will do a dog far less harm than running around on his own finding ways of amusing himself. The average Collie has just as good a nose as the average gun dog and working him as a Spaniel cannot do him any harm as a sheepdog. Provided he is *trained to work properly*, it may do him a lot of good.

As I explained earlier a 'strong-eyed' dog will set game or rabbits like a Setter or a Spaniel. He can be encouraged to walk on exactly as he would to a sheep but on no account must he be allowed to chase anything which he flushes. After all, the same applies to any well-trained gun dog, although by no means to all gun dogs! He can be taught to retrieve when you tell him to retrieve, not to run in when he feels inclined and he will put the average gun dog to shame when it comes to tracking a runner.

Controlled work of this or any other sort will make a dog more biddable and amenable, whether he is being used as a sheepdog, gun dog or just lying around the house. It is essential that the dog is controlled. Nothing ruins a sheepdog or gun dog more quickly than allowing him to hunt rabbits or, worse still, hares willy nilly.

One of the problems which may confront some of my readers is that the farm dog is often expected to work for several different people. This is not an ideal state of affairs but, if the dog is to be a practical asset, is often inevitable. Many excellent dogs are completely ruined by different people using different commands.

There is, of course, no reason at all why different people should not use the same commands in the same way. No reason, that is, except that people are often far more pig-headed than dogs! As far as possible, a young dog should be taught what to do by one person, but once he knows it, he should work just as well for another person he knows and who uses the same commands. That is one reason why I try to keep the commands to a minimum. It makes less for the dog to learn and less for anyone who is not in the habit of working dogs to remember. For instance, if you say 'S-sh' and the dog runs round the cows on either side and brings them up to the gate, he has done the right thing. But if you say 'Come bye' and the dog runs to the right he has done wrong. It must, therefore, be impressed on anyone working the dog not to give commands unless he or she knows what they mean and is able to see they are carried out. If you are being guided by this book in training a young dog it would be a good idea if you could persuade any members of your family or staff who are likely to work the dog to read it too. Not because the methods described are the only ones or the best but because it is only fair to the dog that people who work him should employ the same general principles.

If you want to make a good job of training a young dog it is a great help to practise working an older *trained* one. The novice training a young dog from scratch is like one learning to ride on an unschooled horse. For the few who succeed, many more fall by the wayside!

As far as possible training and working should be regarded as separate operations. Never ask a dog to do something unless you are in a position to see that he does it. That is not always possible in the course of the working day. Training should, therefore, be done quietly when you have time to spare.

This does not mean that a dog cannot be learning while he is working. The fully trained dog is like the person who knows everything – non-existent. A dog can go on learning all his life – if there is someone willing to teach him. Don't do as I, in my inexperienced youth, did with Garry (p. 128–9), ask your dog to do something which you know, or should know perfectly well, he cannot or will not do. Better, far better, not to give a command than give one that is disobeyed.

Be on the lookout for bad habits and stamp on them as they appear but don't forget to encourage good ones. The dog that will go on ahead and have the cows waiting at the gate by the time you reach the field will save more man hours than one that has to be taken there and instructed what to do.

To end *The Farmer's Dog* I am going to repeat a story told in my previous book, *The Family Dog: Its Choice and Training*, as it is even more relevant here. When I left college and started working as shepherd and cattleman on my father's farm, he gave me a young Border Collie bitch, Floss.

Since I was a child I had taught dogs to do tricks, had used Terriers as ratters, etc., and had worked dogs on both sheep and cattle, but had never broken one from scratch. This one was, thanks to my father's attempts at breaking her, well behind scratch, for she had developed some very bad habits. However, a shepherd who gave me much valuable advice on both sheep and sheepdogs assured me that she was a 'giud yin' and would be worth all the trouble. He also made some very unflattering remarks about the ability of my 'old man' and farmers in general to handle dogs, which I kept discreetly to myself!

About a year later another shepherd called. As it happened I wanted some cattle from a forty-acre field some distance from the 'steading' which stood on top of a hill. This field could be reached by a path through a wood, but the cattle had to be taken away to the far end of the field, where there was a gate, on to the farm road. As someone chanced to be going out in the car I asked him to open the gate and sent Floss for the cattle. While I busied myself with something else she went, without further command, through the wood, gathered the cattle in the field, took them to the far end, through the gate, up the road and put them safely in the 'reed', where she lay at the gate until I appeared; and where she would have stayed all night had I not appeared.

Unknown to me this old shepherd had been watching all this and, when I went to shut the cattle in, he came up to me and said, 'That's a good bitch you've got there, lad.' To me, of course, he was only stating the obvious, and his remark only added to the pride I already had.

I then pointed out that this was the first dog I had ever broken and asked if I had not made a good job of it. The reply was neither what I expected nor what I hoped for. He looked straight at me and said in his soft Highland brogue: 'You know, when I was a lad my father gave me a dog and I broke him myself, and he was a topper. I thought I had been very clever. ... ' He paused, then went on, 'But that was nearly sixty years ago and I have bred the same strain ever since – that's one there,' pointing to the dog he had with him, 'but I've *never* had another like him.' I said nothing, but I never forgot those words. It is now just about sixty years since I was told that – during which I have had more dogs through my hands than at that time I could have believed possible – but I have *never* had another like Floss. For a long time I thought and hoped I might – I kept several of her daughters, but none were like her. Experience has taught me that I shall never have another Floss, because no two dogs are alike.

Of course, Floss was not perfect and I do not expect the advice I have given will help you find a *perfect* dog. I hope, however, that I may have helped some farmers towards having an animal that can save them valuable time and money. Even more do I hope that I shall save farm stock some unnecessary harrying about by untrained dogs, not to mention untrained humans! Perhaps, most of all, I hope that I may have improved the lot of many dogs.

No animal, certainly no person, is willing to do so much in return for so little as the farmer's or shepherd's dog. There is no National Union of Agricultural Dogs to argue their case with the N.F.U. Remember, therefore, that your dog not only belongs to you, he depends on you for everything. All he wants is a comfortable bed, some decent food and a master who understands him. Surely that is not too much to give in return for all he does for you. And if he does not do exactly what you want, make sure, before blaming him, that you are not to blame yourself.

And remember that whether he be an outstanding dog or just an ordinary one you will never have another like him – *Never!*

INDEX

head, shape of, 66–7
health, 70–2, 107
hearing, 20–1
heeling, 30, 42
herding instincts, 17–21, 27, 29–31, 47
hill-worn dogs, 43
holding a single, 149–50
house-training, 26, 112–15
housing, 100–7
Huntaways, 64–5
hunting instincts, 17–21, 28, 29

illness, 70–2, 107
instincts, cleanliness, 26–7; development of, 27–32; fear, 24–5, 32; herding, 17–21, 27, 29–31, 47; hunting, 17–21, 28, 29; pack, 21–3, 28; sexual, 25–6, 70–71; submissive, 23, 27, 29–30, 33, 37, 84, 93, 100
intelligence, 28, 36–8, 67, 76–8
International Sheepdog Association, 50, 52–3, 57, 58, 62

Jason, 84–5, 89
Judy, 21, 41, 59, 150, 154

Kelpie, 49, 52, 65–6
Kennel Club, 51, 62
kennels, 26, 103–7, 111

Lancashire Heelers, 61
laziness, 38
leaders, pack, 22–3, 33, 119
leads, 123, 125
lie down, 125–6
lifting, 131–5, 146–7
line, training on, 142–3
loose-eyed dogs, 19, 60, 134

meat, 96
Most, Konrad, 23

Nell, 63–4

old dogs, working young dogs with, 28–9, 129–30

Old English Bobtail, 60, 61, 64
Old Welsh Grey, 60

pack instincts, 21–3, 28
parasites, 71–2
pets, Border Collies as, 51, 52
playpens, 108, 110
problem dogs, 36
punishment, 31–2, 82
puppies, buying 44–8; chewing, 121–3; collars and leads, 123; coming when called, 115–20; diseases, 72; feeding, 94, 99; house training, 26, 112–15; instinct development, 27–32; settling down, 108–12; starting work, 128–9

Quiz, 101–2

ratting, 154
reward, 80, 83–8, 93, 94, 100–1, 124
running, to both hands, 135–7; training, 127–30, 131; wide-running, 43, 53, 54–5, 128
runs, dogs, 103–4

Scotch Collie, 60
sexual instincts, 25–6, 70–71
shedding, 148–9
sheep, mentality, 91–3
show dogs, 51–2
shyness, 34, 47
'sides', 140–3
skin trouble, 72–3
smell, sense of, 20–1
soft dogs, 34–5
South Welsh creeper, 49
spaying, 26
spoiling dogs, 101–3
staying, 126–7
strong-eyed dogs, 19–20, 42, 47, 53, 54, 58, 133, 134, 151
submissive instincts, 23, 27, 29–30, 33, 37, 84, 93, 100

temperament, 33–6, 47, 67, 110
Tessa, 89, 90